Ernst Cassirer on Form and Technology

Ernst Cassirer on Form and Technology

Contemporary Readings

Edited by

Aud Sissel Hoel and Ingvild Folkvord
Norwegian University of Science and Technology, Trondheim, Norway

First published 2012 by
PALGRAVE MACMILLAN

Palgrave Macmillan in the UK is an imprint of Macmillan Publishers Limited, registered in England, company number 785998, of Houndmills, Basingstoke, Hampshire RG21 6XS.

Palgrave Macmillan in the US is a division of St Martin's Press LLC, 175 Fifth Avenue, New York, NY 10010.

Palgrave Macmillan is the global academic imprint of the above companies and has companies and representatives throughout the world.

Palgrave® and Macmillan® are registered trademarks in the United States, the United Kingdom, Europe and other countries.

ISBN 978-0-230-36547-6

This book is printed on paper suitable for recycling and made from fully managed and sustained forest sources. Logging, pulping and manufacturing processes are expected to conform to the environmental regulations of the country of origin.

A catalogue record for this book is available from the British Library.

A catalog record for this book is available from the Library of Congress.

10   9   8   7   6   5   4   3   2   1
21 20 19 18 17 16 15 14 13 12

In memory of John Michael Krois, whose dedication and scholarship inspired this volume

Contents

List of Figures   x

Acknowledgements   xi

Notes on the Contributors   xiii

Introduction   1
Aud Sissel Hoel and Ingvild Folkvord

PART I   FORM AND TECHNOLOGY

1 Form and Technology   15
   Ernst Cassirer (translated by Wilson McClelland Dunlavey and
   John Michael Krois)
   Section I   15
   Section II   22
   Section III   34
   Section IV   42

2 The Age of Complete Mechanization   54
   John Michael Krois

PART II   CONTEMPORARY READINGS

3 Technics of Thinking   65
   Aud Sissel Hoel
   Reframing the problem of technology   67
   Expanding and transforming *logos*   70
   Bios and *logos*   71
   Truth, distance and intervention   74
   Poetic infinity   78
   Technology and the possible   80
   Science criticism   81
   Extended mind   84
   Transformational realism   85

4 The Struggle of Titans – Ernst Jünger and Ernst Cassirer:
   Vitalist and Enlightenment Philosophies of Technology
   in Weimar Germany   92
   Frederik Stjernfelt
   Technics as Titanic destiny   93

Technology as symbolic form 98
Cassirer versus vitalist criticism of technology 102

5 **Technology as Destiny in Cassirer and Heidegger: Continuing the Davos Debate** 113
Hans Ruin
The Davos debate 114
The philosophical challenge of technology 117
Cassirer's 'Form and Technology' 120
Heidegger's initial approach to the problem of technology 125
Heidegger's later approach to the question of technology 130
Concluding comparative remarks 132

6 **Technical Activity as a Symbolic Form: Comparing Money and Language** 139
Jean Lassègue
A disparity in Cassirer's philosophy 139
Three points to reconsider 143
The analogy between money and language 151
Conclusion 157

7 **The Power of Voice: Ernst Cassirer and Bertolt Brecht on Technology, Expressivity and Democracy** 161
Ingvild Folkvord
Hardened scepticism 162
Brecht versus Cassirer 166
Brecht and Cassirer: addressing posterity? 176

8 **'Representation' and 'Presence' in the Philosophy of Ernst Cassirer** 181
Marion Lauschke (translated by Wilson McClelland Dunlavey)
Section I 182
Section II 186
Section III 188

9 **Cultural Poetics and the Politics of Literature** 199
Frederik Tygstrup and Isak Winkel Holm (translated by Lise Utne)
Truth and facts 199
Symbolic forms 202
The politics of literature 206
Creation, exposition, transposition 208

10 **Cave Art as Symbolic Form** 214
Mats Rosengren
Outside 214
Entrance: the study of cave art 214

Hallway and passages: Cassirer on tools and technologies   219
Galleries: on the remaking of the horses in Pech Merle   222
Exit: on the way out of an impasse?   226
Outside again   228

11 **Failures of Convergence**   **233**
Dennis M. Weiss
Section I   233
Section II   238
Section III   242
Section IV   249

Index   259

List of Figures

2.1   22 August 1930: Einstein speaks on the radio to open the
      Radio Exhibition in Berlin (Deutsches Rundfunkarchiv,
      http://www.dra.de/)   57

2.2   21 and 28 October 1931: Cassirer lectures on national
      radio on 'The Unity of Science' over the Deutschlandsender
      near Berlin in their broadcast 'Hochschulfunk'
      (RadioMuseum Köln, http://www.radiomuseum-koeln.de/)   58

7.1   Brecht's radio experiment performed on stage, at the
      music festival in Baden-Baden, 1929 (Akademie der
      Künste, Bertolt Brecht-archiv, http://www.adk.de/)   171

10.1 The Panel of the Horses, Pech Merle, France. © Michel
      Lorblanchet (with permission from the author Michel
      Lorblanchet, *Höhlenmalerei: Ein Handbuch* (Sigmaringen:
      Thorbecke Verlag), 181)   223

10.2 The phases of construction of the Panel of the Horses.
      © Michel Lorblanchet (with permission from the author
      Michel Lorblanchet, 'Rencontres avec le chamanisme',
      in Michel Lorblanchet, Jean-Loïc Le Quellec, Paul Bahn
      and Henri-Paul Francfort (eds), *Chamanisme et Arts
      Préhistoriques: Vision Critique* (Paris: Errance, 2006),
      105–36, 107)   225

Acknowledgements

This book is dedicated to our dear friend and colleague John Michael Krois, who was tremendously supportive of and served almost as a mentor for our work on Ernst Cassirer's 'Form and Technology'. We came to know John in 2006 when we invited him to Trondheim to give a keynote presentation at our conference *Form and Technics: Reading Ernst Cassirer from the Present*. After that we met John on several occasions in Berlin and Paris. We fondly remember our trips to Berlin, where John spontaneously took on the role of the indigenous guide, eager to share the hidden attractions of the city that he loved so much. Until his sudden illness and death in the autumn of 2010, John contributed to our scholarly work in many ways. With respect to this volume, John's effort can be identified through his co-translation of the volume's topical essay – Cassirer's 'Form and Technology' – and through his own essay 'The Age of Complete Mechanization', which contextualizes Cassirer's classic text. Far beyond that, John contributed with his acumen as a philosopher, his generous sharing of his vast knowledge of Cassirer's life and work, his relentless support and, not the least, his various ways of bringing together people with shared intellectual interests. Our first thanks therefore go to John.

We would also like to express our thanks to the other contributors to this volume, several of whom were present at a workshop that we organized at the Nordeuropa-Institut at the Humboldt University in Berlin in 2008. Thank you for your patience! Thanks also to Bernd Henningsen, who hosted our 2008 workshop and who, together with John, invited us to participate in the annual Cassirer workshops at the Nordeuropa-Institut. We want to express our thanks to the Faculty of Humanities at the Norwegian University of Science and Technology (NTNU) for financially supporting, among other things, our 2006 conference and our 2008 workshop, and to the NTNU Departments of Modern Foreign Languages and Art and Media Studies, for financially supporting translation and proofreading work.

The repository of the Ernst Cassirer Papers, including the essay 'Form und Technik', is the Beinecke Rare Book and Manuscript Library, Yale University. The editors and publishers are grateful to Yale University Press for the kind permission to use and translate this essay for this volume; to Deutsches Rundfunkarchiv, to the RadioMuseum Köln; to the Bertolt Brecht-Archiv at Akademie der Künste for permission for photographs; to Michel Lorblanchet for his photograph and drawings of the Panel of the Horses; and to Biblioteca Nacional de España for the drawing by Leonardo da Vinci on the front jacket of the book.

Notes on the Contributors

Wilson McClelland Dunlavey is a Friedrich Naumann Scholar at the Humboldt University of Berlin. He studied Ernst Cassirer with John Michael Krois for six years and is currently writing his doctoral dissertation on the influence of German cultural programmes on American universities. He has written several articles on philosophy and history, including 'Vivre c'est essayer: Montaignes Philosophie heißt Sterben lernen' in *Was ist Leben: Festgabe für Volker Gerhardt* (2009) and 'The Pilgrimage to Weimar: Goethe and the American Transcendentalists' (forthcoming).

Ingvild Folkvord is Associate Professor of German literature at the Norwegian University of Science and Technology. She is the author of *Sich ein Haus schreiben: Drei Texte aus Ingeborg Bachmanns Prosa* (2003) and has written several articles on modern literature, radio culture and literacy. She is the co-editor of *Rammer for skriving* (2009), *Skriving i kunnskapssamfunnet* (2010) and *Ernst Cassirer: Form og teknikk: Utvalgte tekster* (2006) and has translated Ingeborg Bachmann ('Undine geht', in *Korttekster* (2000)) and Ernst Cassirer (*Form og teknikk: Utvalgte tekster* (2006)).

Aud Sissel Hoel is Associate Professor of Visual Communication at the Norwegian University of Science and Technology. She is the author of *Maktens bilder* (2007) and *Fremstilling og teknikk: Om bildet som formativt medium* (2005); co-editor of *Computational Picturing*, a special issue of *Interdisciplinary Science Reviews* (2012), and *Ernst Cassirer: Form og teknikk: Utvalgte tekster* (2006); and author of articles on photography, scientific imaging, and the philosophy of vision.

Isak Winkel Holm is Associate Professor of Comparative Literature at the University of Copenhagen. He is the author of *Tanken i billedet: Søren Kierkegaards poetik* (1998) and articles on Rousseau, Schlegel, Kleist, Hegel, Nietzsche, Dostoevsky, Musil, Kafka, Kundera, DeLillo, Sebald and McCarthy. He has translated Franz Kafka, *Fortællinger* and *Efterladte fortællinger* (2008), and Friedrich Nietzsche, *Tragediens fødsel* (1996).

John Michael Krois was First Senior Professor of Philosophy at the Humboldt University of Berlin, Germany. He is the author of *Körperbilder und Bildschemata: Aufsätze zur Verkörperungstheorie ikonischer*

Formen (published posthumously 2011) and *Cassirer: Symbolic Forms and History* (1987); co-editor of *Edgar Wind: Kunsthistoriker und Philosoph* (1998); co-editor of Ernst Cassirer's *Nachgelassene Manuskripte und Texte* (1995–2009); co-editor and translator of Cassirer, *The Philosophy of Symbolic Forms*, vol. 4, *The Metaphysics of Symbolic Forms* (1996); and co-editor of *Symbolic Forms and Cultural Studies: Ernst Cassirer's Theory of Culture* (2004).

Jean Lassègue is a philosopher attached to the French Scientific Research Council (CNRS). He is a member of the CREA (Centre de Recherche en Epistémologie Appliquée) at École polytechnique, Paris. He is the author of several books on the emergence of computer science, from both an epistemological and an anthropological point of view. His interest in Cassirer is rooted in the notion of a symbolic form which he used as a general operator to describe symbolic activities.

Marion Lauschke is a research associate at the Collegium for the Advanced Study of Picture Act and Embodiment, Humboldt University of Berlin. She is the author of 'Das Erhabene bei Ernst Cassirer: Scheitern des Synthesevermögens oder Kontinuum des Formbegehrens?', in *Philosophie der Kultur: Kultur des Philosophierens Ernst Cassirer im 20. und 21. Jahrhundert* (2012) and 'Les fonctions de l'art chez Ernst Cassirer', in *Ernst Cassirer et l'art comme form symbolique* (2010); editor of *Ernst Cassirer: Schriften zur Philosophie der symbolischen Formen* (2009); and author of *Ästhetik im Zeichen des Menschen. Die ästhetische Vorgeschichte der Symbolphilosophie Ernst Cassirer und die symbolische Form der Kunst* (2007).

Mats Rosengren is Professor of Rhetoric at Södertörn University, a member of the editorial board of the philosophical journal *Glänta* and president of the Swedish Ernst Cassirer Society (www.glanta. org/cassirer). He is the author of 'On Creation, Cave Art and Perception: A Doxological Approach', *Thesis Eleven* (2007), *De symboliska formernas praktiker: Ernst Cassirers samtida tänkande* (2010), *För en dödlig, som ni vet, är största faran säkerhet: Doxologiska essäer* (2006) and *Doxologi: En essä om kunskap* (2002/2008, in French as *Doxologie: Essai sur la connaissance* (2011)). He is the editor of *Politics of Magma* (2008) and co-editor of *Embodiment Rediscovered* (2007).

Hans Ruin is Professor of Philosophy at Södertörn University in Stockholm. He is co-editor of *Fenomenologi, teknik och medialitet* (2011) and *New Frontiers: Phenomenology and Religion* (2010); author of 'Ge-stell' in *Martin Heidegger Basic Concepts* (2009); co-editor of *The*

Past's Present: Essays in the Historicity of Philosophical Thought (2006); author of *En Kommentar till Heideggers Varat och tiden* (2005); co-editor of *Metaphysics, Facticity and Interpretation* (2003) and of Nietzsche's *Collected Works* in Swedish.

Frederik Stjernfelt is Full Professor at the Centre for Semiotics, Aarhus University, Denmark. He is the author of papers in international and Danish journals and some ten books in English, Danish, Norwegian, Swedish, Serbian – most recently *Diagrammatology* (2007) and *Adskillelsens politik* (2008, with Jens-Martin Eriksen; French and English versions to appear in 2012). He is co-editor, with Peer Bundgaard, of *Semiotics: Critical Concepts* (4 volumes, 2010). His research interests include semiotics, cognitive science, epistemology, history of ideas, theory of literature and political philosophy.

Frederik Tygstrup is Director of the Copenhagen Doctoral School in Cultural Studies and an associate professor of comparative literature at the University of Copenhagen. He is the author of *Erfaringens fiktion: Essay om romanens form* (1992) and *På sporet af virkeligheden* (2000); co-editor of *Illness in Context* (2010) and *Witness: Memory, Representation, and the Media in Question* (2008); and author of some 80 articles in Danish, English, French and German on a wide range of topics in comparative literature and cultural studies. His present research interests focus on the intersections of artistic practices and other social practices, including urban aesthetics, the history of representations and experiences of space, literature and medicine, literature and geography, literature and politics.

Dennis M. Weiss is Professor of Philosophy at York College of Pennsylvania. He is the editor of *Interpreting Man* (2003); author of 'Transforming the Symbolic Animal: Ernst Cassirer and the Posthuman', *Humanities and Technology Review* (2011); co-author, with Rebecca Kukla, of '"The Natural Look": Extreme Makeovers and the Limits of Self-Fashioning', in *Cosmetic Surgery: A Feminist Primer* (2009); author of 'Human-Technology-World', *Techne* (2008) and 'Humanity at the Turning Point: Philosophical Anthropology and the Posthuman', *Expositions* (2007). His present research interests focus on the intersection of philosophical anthropology and the philosophy of technology, especially as they converge on the theme of the post-human.

Introduction

Aud Sissel Hoel and Ingvild Folkvord

There seems to be a Cassirer revival going on. In Germany, especially, the number of dissertations treating Ernst Cassirer's philosophy, and his philosophy of symbolic forms in particular, is increasing every year. The same trend can be seen in the Anglophone world, where new books on and new translations of Cassirer are coming out every year. Why this renewed interest, one is tempted to ask, and why now? Certainly, there are historical reasons. More and more people have come to realize that Ernst Cassirer – a distinguished philosopher of the German idealist tradition, admirer of Johann Wolfgang von Goethe, pronounced supporter of the Weimar Republic and a cosmopolitan liberal of Jewish background who at the height of his career had to leave his position and flee the Nazis – has not received the attention he rightfully deserves. What incites the present revival, however, is not merely an urge to raise a monument to a great thinker. It is spurred, rather, by pressing current concerns, such as the vacuum left by the receding paradigm of poststructuralism in the cultural sciences, or by the onslaught, across disciplines, of new reductive biologisms in the wake of the recent proliferation of evolutionary psychology and related gene-centred approaches. Furthermore, it is prompted by the way that Cassirer's philosophy of symbolic forms provides rich and still untapped resources for the ongoing attempts to bridge unproductive intellectual gaps. Cassirer's thinking is unique in the way that it endeavours to integrate logical concerns, championed by scientifically oriented philosophers, with the concerns of the historical and cultural sciences. Standing, as he did, at the threshold of what has come to be an ever-widening gulf separating 'rationally' inclined analytic philosophers from 'irrationally' inclined continental philosophers, and whose symbolic inception was marked by the legendary meeting between Cassirer and Martin Heidegger at the

1

Swiss resort of Davos in April 1929,[1] Cassirer made a serious effort to negotiate these, by now, all too familiar intellectual tensions.

The key to this sought-after integration is the concept of 'symbolic function', which, according to Cassirer, has validity across the entire domain of human meaning-making, ranging from perception, via language and art, to mathematics. What characterizes human existence is that it is essentially mediated by what Cassirer refers to as 'symbolic forms'. The prime examples of symbolic forms are language, myth and science, each of which is devoted a volume in Cassirer's three-volume main exposé of his philosophy of symbolic forms (1923, 1925, 1929).[2] Other symbolic forms include art, religion and history. Even if they are historical and variable, the mediating forms are not conceived in opposition to nature. They are not understood to abolish or uproot the determinations that human beings are subjected to by virtue of being living organisms. Instead, the mediating forms are understood to introduce a peculiar *distance* into the living nexus of action and reaction, which brings about a change in the meaning and function of these determinations. This distance, characteristic of all human conduct and meaning-making, grants to human existence a peculiar *leeway* that Cassirer describes in terms of a relative freedom. It is worth noticing that, on the latter account, the distancing work performed by mediating forms is not understood to alienate humans from nature or reality, or from some supposed core of authentic life behind or before the circle of symbols. The distance introduced by symbolic intermediaries is understood, rather, to allow humans to grasp reality in new and more profound ways: 'The individual "symbolic forms" ... are the specific media that man has created in order to separate himself from the world through them, and in this very separation bind himself all the closer to it.'[3] It is precisely here, in his conception of mediation as a dynamic, double and two-way process, that Cassirer's philosophy breaks new ground and opens the way for what has been characterized as a 'radical middle road';[4] that is, as a genuine third position, between relativism on the one hand and objectivism on the other.

Symbols and signs figure prominently in Cassirer's writings. It is important to note, however, that for Cassirer symbols are not representational in the Cartesian sense, nor are they arbitrary in the sense of contemporary semiology and poststructuralism. The symbolic forms do not relate to a pre-given reality; nor do they stand in for or supplement anything. They are considered instead in terms of their *achievements*, of what they *do*. Cassirer's theory of symbols or mediation is directed, so to speak, forwards rather than backwards, in that it combines receptivity with a strong productive impulse. The symbolic forms are conceived of as *tools*, and the power of tools resides precisely in their capacity to incite changes. This is to say that the philosophy of symbolic forms *diverges* from most established theories of signs and meaning in that it proceeds on a *positive notion of difference*.[5] The transformations induced by the mediation process are not understood to jeopardize or postpone the identity of the object. Quite the contrary, it is merely by virtue of an intervening medium that phenomena come into view as objects in the first place, as entities that are granted some kind of independent existence. This is because the symbolic intermediary provides the necessary ingredient in any identifying, measuring or comparative act: a *reference point* – or, what for Cassirer amounts to the same, a *viewpoint*.

Cassirer's philosophy treats symbols on different levels. A name, for instance, may serve as a point of crystallization ascribing a certain order and unity to phenomena.[6] When he sets out to define the concept of symbolic form, however, Cassirer provides us with a list of very general systems of human activities, which typically includes the examples already mentioned (language, myth, science, art, religion, history). Despite the fact that he conceives of symbolic forms as tools, it is not until the essay made topical in this volume, 'Form and Technology',[7] published in 1930, that he devotes serious attention to the problem of technology. Cassirer's treatment of technology, therefore, appears as something of an afterthought, delivered after he had completed the third and final volume on the philosophy of symbolic forms. What becomes clear, however, even from a superficial reading of the technology essay, is that Cassirer now makes a considerable investment in the new topic. He is not content merely to include technology as yet another system of human activities alongside the others already on the list. The essay addresses far more fundamental questions concerning the tool character of language and the theoretical work performed by material instruments, proceeding to advance technology as a primary medium in equal company with language and art.

Viewed against the background of the ambitious programme for a new kind of philosophy of technology that germinated in this essay, and that ascribes to technology a new dignity as a genuine tool of the mind, some of the criticisms that have been raised against Cassirer's concept of technology simply seem off the mark. Taking into consideration what the essay says about the inner relations between theory and material instruments, we cannot agree with Johannes Rohbeck when he claims that Cassirer advances a concept of technology that is 'filtered and purified'.[8] Certainly, the essay does not delve into examples, and in a longer exposition a treatment of particular technologies would be called for. Yet, what Cassirer aims for in this essay is a change of track, a radical change in viewpoint that allows the question concerning technology to be posed in an altogether new way. He seeks to establish a fresh angle of approach that allows the inner relations between technology, language and art to come to the fore and thus to be investigated. Likewise, considering the above discussion of the tool character of symbols, we cannot agree with Gideon Freudenthal when he claims that Cassirer's theoretical approach is characterized by a 'disregard for technology and for tool use in general'.[9] As we see it, the strength of Cassirer's approach has precisely to do with the way that it challenges and undermines the long-established opposition between theory and practice. In sharp contradistinction to the standard Aristotelian view, 'theoria', for Cassirer, is not conceived in opposition to 'praxis' or 'techne'. Nor do these terms refer to activities that can be performed separately. What allows for the new inner relation between theory and practice is the weight accorded to the dynamic notions of 'formation' and 'work' (production). It is no coincidence that Cassirer's writings are permeated by metaphors of building and constructing. Theory is conceived as a tool-based activity that is performed as a dynamic cultural practice, as agency: concepts and notations need to be forged and coined, theories worked out and demonstrated, models built and arguments need to be set forth and defended. This is to say that theoretical work is a situated practice that addresses a specific community and plays out in a shared yet highly specialized public sphere, and whose precarious constructions are exposed to the constant scrutiny of fellow researchers.

It is possible to observe how Cassirer's insights manifest themselves, not merely as theoretical claims but also on the micro-level of Cassirer's characteristic style of writing. As if to account for a pragmatically oriented material hermeneutics, he makes use of textual markers that emphasize the productive and performative aspects of theoretical work. In 'Form and Technology', for example, when he exposes the intellectual contributions made by previous scholars, he frequently uses significant combinations of verbs: he discusses what Wilhelm von Humboldt has 'said and proven' and what Theodor-Wilhelm Danzel has 'maintained and carried through'.[10] A further peculiarity of his writing concerns the way that Cassirer presents the views of his theoretical and political adversaries. True to his principle of theoretical reflection through performative display, he typically articulates their views and positions in an elaborate fashion, establishing them in their profundity and plenitude. His method of argument consists of going along with the view under discussion as far as he can, seemingly endorsing it. It is only at the end that he marks the point of deviation, critiques his adversaries' most basic premises and positions himself in contra-distinction to the presented view.[11] This sympathetic strategy, which he also undertakes in 'Form and Technology', may sometimes cause confusion, since the careful expositions of the views held by theoreti-cal adversaries may easily be mistaken for Cassirer's own views. Again, there is an interesting link between the performative aspects and the insights professed. Cassirer accords great importance to the activity he refers to as 'Auseinandersetzung' – a term for which most English trans-lations are negative (opposition, struggle, conflict, antagonism), but which for Cassirer designates the productive act of determining some-thing through processes of differentiation that bring about a revealing contrast and counter-positioning. In fact, on the latter account, clas-sification is conceived as a dynamic process that develops through the comparison of divergent points of view.

Even though Cassirer in 1930 makes a strong effort to integrate tech-nology into his general philosophical framework, 14 years later, when he summarizes his position for the English-speaking community in *An Essay on Man* (1944), he is back on the old list.[12] This observation prompts us to raise the question, this time going along with Rohbeck's critical remarks, regarding the position of 'Form and Technology' in rela-tion to Cassirer's main contribution, the philosophy of symbolic forms. If technology is indeed a primary medium, as the essay suggests, would not this insight – so rich in implications – provoke a radical change in Cassirer's main doctrine as it appears in its canonical three-volume instantiation?[13] Would not this insight increase the tensions already present in the original doctrine, concerning, say, the status of myth or language vis-à-vis the other symbolic forms? Or the status of mathemat-ics, which Cassirer continues to privilege as the supreme expression of the human intellect even after his philosophy has taken a cultural turn? Cassirer did not pursue the implications and ramifications of the insights set forth in the 1930 essay. Seen against the overall picture of his philosophical output, technology remains an add-on – perhaps even a dangerous supplement in the sense of Jacques Derrida.[14] What is more, if we compare the views on technology presented in 'Form and Technology' with the views presented in his last work, *The Myth of the State* (published posthumously in 1946), we may easily be led to conclude that Cassirer, after having witnessed the atrocities of the Third Reich, feels compelled to go back on his mainly positive stance on modern technology presented in the 1930 essay. For, at least on the face of it, the views presented in the 1946 work seem to be more in line with the negative stance associated with Heidegger's analysis of modern technology and the Frankfurt School.[15] Such a conclusion, however, would be hasty and inconsiderate on at least two accounts. First, even if both texts are concerned with technology, they differ in their overall topics and objectives. While 'Form and Technology' undertakes a systematic reflection of technology's potential as a symbolic form, *The Myth of the State* investigates not technology as such, but rather the dangers involved when myth is enrolled in the service of state power and used as a social technique of political persuasion and control.[16] Second, 'Form and Technology' already points to the dangers associated with what Cassirer sees as a conflict between symbolic forms, where 'each form threatens to subjugate all other forms to itself'.[17] However, it is not until *The Myth of the State* that Cassirer pursues the social and political implications of this conflict between forms.

In any event, these critical remarks and contrasting interpretations do not compromise the philosophical significance of 'Form and Technology'. To the contrary, they serve rather to heighten the interest of this highly original and thought-provoking essay, where Cassirer sets out to answer the charges directed against modern technology. Moreover, Cassirer was not a dogmatic thinker, and he did not consider the philosophy of symbolic forms a finished system. He saw it, rather, as a prolegomena to a future philosophy of culture, welcoming critical revisions.[18] This leaves much room for present-day thinkers who want to trace the hitherto unattended-to connections hinted at in the technology essay or to rework the philosophy of symbolic form in such a way that it can meet the theoretical concerns of the current day. This, at least, is the spirit in which we want to put forward Cassirer's essay on technology in the present volume, where it is offered for the perusal of the English-speaking world for the first time.[19]

The present volume is divided into two parts. Part I provides an English version of 'Form und Technik', which has been translated by Wilson McClelland Dunlavey and John Michael Krois. The first part also includes a short essay by Krois, 'The Age of Complete Mechanization', which situates the topical essay in its historical and cultural context. In his essay, Krois draws a picture of an epoch – 1920s Germany – where technology has become a matter of general concern for everyone: intellectuals, artists, politicians and the general public. Modern technology is perceived as a threat to humanity as well as its saviour, and the perils and prospects of complete mechanization are subjected to excited artistic and philosophical treatment. The essay also treats the role played by Leo Kestenberg, the editor of the volume where Cassirer's technology essay first appeared, as well as the importance of radio to the cultural situation, and of Cassirer and his family's involvement in radio broadcasting.

The second part of the volume consists of nine critical essays that discuss the philosophical significance of Cassirer's essay on technology. All contributors to the volume have been assigned the same task: to account for the current relevance of Cassirer's philosophy of symbolic forms, and in particular, his approach to technology, with a view to issues of concern in their respective disciplines.

Aud Sissel Hoel's essay, 'Technics of Thinking', explores the potential of Cassirer's notion of symbolic and, especially, technological mediation, which allows for a fresh take on the old problems of knowledge and mind. Building on Cassirer's core insight, delivered in the technology essay, concerning how tools ground the sort of mediacy that makes thinking possible, she sketches what she describes as a 'differential' approach to knowledge. The resulting approach is at variance in significant respects from the philosophy of difference associated with poststructuralist thinking, and Hoel goes on to compare it, instead, with lines of thinking pursued in the contemporary philosophy of technology, in science and technology studies and in recent 'extended mind' approaches within the philosophy of mind.

In his essay, 'The Struggle of Titans – Ernst Jünger and Ernst Cassirer: Vitalist and Enlightenment Philosophies of Technology in Weimar Germany', Frederik Stjernfelt compares Ernst Cassirer's philosophy of technology, as presented in 'Form and Technology', with the views on technology endorsed by Cassirer's contemporary, Ernst Jünger, as presented mainly in the book-length essay *The Worker* (1932).[20] Whereas the former sees technology as one device among many that allow human beings to interact with and construct the world, and thus advancing an optimist neo-Kantian interpretation of technology, the latter sees technology as a destiny that mankind cannot avoid and that must be pushed to its end in order for new values to be shaped, and thus instigating a Nietzschean and vitalist approach to technology which was to influence Heidegger and other conservative criticisms of technology. By comparing the two, Stjernfelt articulates two very different, almost antagonistic interpretations of technology, which still influence main versions of philosophies of technology, making a strong case for the enlightenment version.

Hans Ruin's 'Technology as Destiny in Cassirer and Heidegger: Continuing the Davos Debate' provides a parallel reading of Cassirer's

'Form and Technology' and of Heidegger's contribution to the philosophy of technology, mainly through the essays 'The Age of the World Picture' (1938)[21] and 'The Question Concerning Technology' (1953).[22] Ruin approaches Heidegger's views as a response to and as an implicit elaboration of Cassirer's approach. He conceives the relation between the two thinkers as a virtual dialogue, the stakes for which were laid down in the debate between the two in the 1929 Davos debate. After providing a detailed account of their respective contributions at Davos, he goes on to reconstruct Cassirer's and Heidegger's views on technology, not only with an eye to their differences but also with an eye to their common ground, which Ruin refers to as their 'shared Kantian matrix': both thinkers accentuate the way technology changes the meaning of nature and being, and both thinkers see technology as a destiny of man in modernity. They differ markedly, however, when it comes to the prospects of developing a rational-ethical critique of the technology-saturated present.

Jean Lassègue, in his essay 'Technical Activity as a Symbolic Form: Comparing Money and Language', develops what it means for technical activity to be understood as a symbolic form and how it differs from other symbolic forms such as language. Whereas Cassirer in 'Form and Technology' sees technical activity as the crucial distinguishing factor between animals and humans, Lassègue re-evaluates this difference. Drawing on recent findings in ethology, primatology and paleo-anthropology, which confirm collective technical activity among animals and pre-human primates, he proposes that animal technical activity differs from human technical activity in that it is not yet symbolic. Human technical activity – technology – is characterized by heterogeneity – that is, by a peculiar connection between instability and stability in its inner features. Thus, it is not technical activity as such, but technology – which has a symbolic dimension – that belongs to the domain of symbolic forms. In like manner, Lassègue goes on to re-evaluate two other topics from 'Form and Technology': the idea of technical activity as organ-projection and the idea of technical activity as an anticipation of self-knowledge. The essay concludes with a comparison between money, conceived as technology, and language.

Ingvild Folkvord explores, in 'The Power of Voice: Ernst Cassirer and Bertolt Brecht on Technology, Expressivity and Democracy', the potential of Cassirer's approach to technology and expressivity by juxtaposing it to Bertolt Brecht's critical reflections on radiophone mediation. In spite of all the obvious differences, the Marxist author and the liberal philosopher meet in their concerned reflections on the effects of new technology on democratic development, a concern that appears to be closely intertwined with the political tensions in the late period of the Weimar Republic in which both Cassirer and Brecht lived and worked. The comparison seeks to demonstrate how Cassirer's systematic recognition of expressive phenomena allows for a dynamic understanding of technically mediated voice phenomena that questions the sceptical approach which has become *doxa* in the contemporary field of cultural studies and aesthetics.

In her essay, '"Representation" and "Presence" in the Philosophy of Ernst Cassirer', Marion Lauschke argues that the philosophy of symbolic forms provides a constructivist theory of representation. Like Kant, Cassirer sees experience in terms of coherencies that follow certain rules of connection, but in contrast to Kant he pluralizes the constitutive relations through which these ordered structures are established. In the first section of the essay, Lauschke expounds on the concept of 'representation' that underlines the philosophy of symbolic forms and contrasts it with the concept of 'presentation'. She goes on to discuss to what extent these two concepts help differentiate the various symbolic forms from each other. In the concluding section, Cassirer's theory of symbols is brought to bear on art and aesthetic phenomena, and compared with contemporary approaches such as Hans Ulrich Gumbrecht's notion of production of presence and Jacques Derrida's deconstructionist concept of signification. In line with Cassirer, Lauschke suggests that aesthetic phenomena are shaped by an oscillation between presence and representation.

'Cultural Poetics and the Politics of Literature', by Frederik Tygstrup and Isak Winkel Holm, uses Ernst Cassirer's concept of 'symbolic forms' to suggest that a distinction be made between a general cultural poetics and a specific literary poetic. Any culture possesses a common repertoire of narratives and cognitive forms that can be used to configure facts. A cultural poetics consists of a set of collective techniques and principles enabling the production of cultural images of reality. The poetics of literature, on its side, is simultaneously part of and apart from this poetics of culture. Due to this duality of likeness and difference, literature takes the guise of a cultural laboratory where the roles and functions of the cultural repertoire of images of reality can be tested. Tygstrup and Holm argue that the political significance of literary representational practices resides in the way they activate a common cultural repertoire of historical symbolic forms while at the same time deviating from the common ways of treating these forms. They conclude by distinguishing among three different modes of literary deviation: the creation of new images of reality, the exposition of a culture's repertoire of images of reality and the transposition of images of reality from one institutional context to another.

Mats Rosengren's 'Cave Art as Symbolic Form' takes as its starting point the lack of consensus displayed concerning the interpretations of Paleolithic cave art, one of the earliest forms of symbolic expression known to us. According to Rosengren, cave art studies are marred by two recurring problems: a ubiquitous longing for an origin and a stubborn conviction that seeing and depiction take on meaning by passively reflecting what is already there – a conviction that he refers to as the 'mimetic curse'. Through its connection of life and thought as embodied in man's tools, Cassirer's philosophy of symbolic forms, and particularly his philosophy of technology, provides a new and more productive framework for the study of cave art. Drawing on Cassirer's characterization of symbolic forms as organs rather than mirrors of reality, and building on the discussion of tools and organ-projection in 'Form and Technology', Rosengren outlines a new and experimental method for making sense of the traces found in the caves. He substantiates his approach by considering a concrete experiment conducted by the French cave art specialist Michel Lorblanchet.

Dennis M. Weiss's essay, 'Failures of Convergence', observes that it is widely suggested today that technology has advanced to the point where it has the capacity to fundamentally transform the conditions of human life. Proponents of what has come to be referred to as nano-bio-info-cogno (NBIC) technologies, predict a soon-to-be future in which these converging technologies will result in improved human performance and a golden age of social development that can transcend current crises. Critically appraising these calls for a convergence of technology on improving human performance, Weiss draws on the formative work of Ernst Cassirer on philosophical anthropology, culture and technology, and argues that a more adequate account of convergence, which addresses the place of the human being in a rapidly changing technological environment, must begin from a stance that incorporates philosophical anthropology and a critical theory of technology. Whereas NBIC convergence provides a framework for the unity of disciplines predicated on nanotechnology, the unity of nature, cause-and-effect thinking and a hierarchy of disciplines, Cassirer resists these kinds of linear, hierarchical and ultimately reductive models. The crises facing human beings cannot be solved by means of technology alone. For Cassirer, progress, far from being a scientific task, is an ethical task and a perpetual one.

Notes

1. For recent discussions of the philosophical significance of the Davos encounter, see Michael Friedman, *A Parting of the Ways: Carnap, Cassirer and Heidegger* (Chicago: Open Court, 2000) and Peter Eli Gordon, *Continental Divide: Heidegger, Cassirer, Davos* (Cambridge: Harvard University Press, 2010).

2. The main exposition of Cassirer's philosophy of symbolic forms was published in three volumes by Bruno Cassirer Verlag: *Philosophie der symbolischen Formen*, vol 1: *Die Sprache* (1923), *Philosophie der symbolischen Formen*, vol. 2: *Das mythische Denken* (1925) and *Philosophie der symbolischen Formen*, vol. 3: *Phänomenologie der Erkenntnis* (1929). English translations were published by Yale University Press: *Philosophy of Symbolic Forms*, vol. 1: *Language* (1953), *Philosophy of Symbolic Forms*, vol. 2: *Mythical Thought* (1955) and *Philosophy of Symbolic Forms*, vol. 3: *The Phenomenology of Knowledge* (1957). Cassirer planned a fourth volume in the series, which was to deal with the metaphysics of the symbolic forms. A set of texts and working notes intended for this project was published in 1995, in a volume edited by John Michael Krois and Donald Phillip Verene and entitled *Nachgelassene Manuskripte und Texte*, vol. 1: *Zur Metaphysik der symbolischen Formen* (Hamburg: Meiner). An English translation of these texts was published in 1996 and entitled *The Philosophy of Symbolic Forms*, vol. 4: *The Metaphysics of Symbolic Forms* (New Haven: Yale University Press).

3. Ernst Cassirer, *The Logic of the Cultural Sciences: Five Studies* (New Haven: Yale University Press, 2000), 25.

4. Ute Daniels, *Kompendium Kulturgeschichte* (Frankfurt am Main: Suhrkamp, 2006), 91.

5. For a discussion of Cassirer's approach to identity and difference, see Aud Sissel Hoel, 'Thinking "Difference" Differently: Cassirer versus Derrida on Symbolic Mediation', *Synthese* 179 (2011), 75–91.

6. Ernst Cassirer, 'Die Sprache und der Aufbau der Gegenstandswelt', in Birgit Recki (ed.), *Gesammelte Werke* (Hamburg: Felix Meiner Verlag, 1998–2009), vol. 18: *Aufsätze und kleine Schriften (1932–1935)* (2004), 119.

7. Original title 'Form und Technik', first published in Leo Kestenberg (ed.), *Kunst und Technik* (Berlin: Wegweiser, 1930), 15–61.

8. Johannes Rohbeck, 'Technik und symbolische Form bei Cassirer', in Peter A. Schmid and Simone Zurbuchen (eds), *Grenzen der kritischen Vernunft: Helmut Holzhey zum 60. Geburtstag* (Basel: Schwabe & CO AG Verlag, 1997), 202.

9. Gideon Freudenthal, 'The Missing Core of Cassirer's Philosophy: Homo Faber in Thin Air', in Cyrus Hamlin and John Michael Krois (eds), *Symbolic Forms and Cultural Studies: Ernst Cassirer's Theory of Culture* (New Haven: Yale University Press, 2004), 218.

10. See Ernst Cassirer, 'Form and Technology' in this volume, 22 and 25.

11. As pointed out by Edward Skidelsky, another method of argument frequently employed by Cassirer is to allow his own position to emerge indirectly through his historical narratives. Edward Skidelsky, *Ernst Cassirer: The Last Philosopher of Culture* (Princeton, NJ: Princeton University Press, 2008), 229.

12. In *An Essay on Man*, he writes: 'Man's outstanding characteristic, his distinguishing mark, is not his metaphysical or physical nature – but his work. It is

this work, it is the system of human activities, which defines and determines the circle of "humanity". Language, myth, religion, art, science, history are the constituents, the various sectors of this circle. A "philosophy of man" would therefore be a philosophy which would give us insight into the fundamental structure of each of these human activities, and which at the same time would enable us to understand them as an organic whole.' Ernst Cassirer, *An Essay on Man* (New Haven: Yale University Press, 1944), 68.

13. Rohbeck, 'Technik und symbolische Form bei Cassirer', 198.
14. A dangerous supplement is a supplement that does not blend seamlessly with a given system, but contains elements that undermine or threaten the integrity of the original system. Jacques Derrida, *Of Grammatology* (Baltimore: Johns Hopkins University Press, 1997).
15. Most notably in Max Horkheimer and Theodor W. Adorno's *Dialektik der Aufklärung* (1947; published in English as *Dialectic of Enlightenment* in 1972) and in Martin Heidegger's 'Die Frage nach der Technik' (1953, lecture, published in English as 'The Question Concerning Technology').
16. For a discussion of these aspects of *The Myth of the State*, see Esther Oluffa Pedersen, *Die Mythosphilosophie Ernst Cassirers* (Würzburg: Königshausen & Neumann, 2009), 206–39.
17. John Michael Krois, 'Ernst Cassirer's Theory of Technology and its Import for Social Philosophy', *Research in Philosophy & Technology*, 5 (1982): 209–22, 215.
18. Ernst Cassirer, 'Zur Logik des Symbolbegriffs', in Ernst Cassirer, *Wesen und Wirkung des Symbolbegriffs* (Darmstadt: Wissenschaftliche Buchgesellschaft, 1956), 229. Originally published in *Theoria* 4 (1938), 145–75.
19. The essay has previously been translated into French ('Forme et technique', trans. Jean Carro and Joël Gaubert, in Ernst Cassirer, *Écrits sur l'art: Forme et technique*, ed. Fabien Capeillères (Paris: Cerf, 1995)); into Norwegian ('Form og teknikk', trans. Ingvild Folkvord, in Ernst Cassirer, *Form og teknikk: Utvalgte tekster*, eds Ingvild Folkvord and Aud Sissel Hoel (Oslo: Cappelen, 2006)); and into Danish ('Form og teknik', trans. Marie Møller Kristensen, in: *Teknologi & virkelighed*, eds Jan Kyrre Berg Olsen and Stig Andur Pedersen (Århus: Philosophia, 2008)).
20. *Der Arbeiter: Herrschaft und Gestalt* (Hamburg: Hanseatische Verlagsanstalt, 1932).
21. 'Die Zeit des Weltbildes', lecture delivered in 1938.
22. 'Die Frage nach der Technik', lecture delivered in 1953.

Part I
Form and Technology

1
Form and Technology[1]

Ernst Cassirer

I

If we judge the significance of the individual areas of human culture primarily by their actual effectiveness, if we determine the value of these areas according to the impact of their direct accomplishments, there can hardly be any doubt that technology claims the first place in the construction of our contemporary culture. Likewise, no matter whether we reproach or praise, exalt or damn this 'primacy of technology', its pure actuality seems to be beyond question. All the formative energy in contemporary culture is increasingly concentrated on this one point. Even the strongest counter-forces to technology, even those intellectual forces that are the most distant from technology in their content and meaning, seem able to actualize themselves only insofar as they become conjoined with technology and, through this alliance, become imperceptibly subjected to it. Today many consider this subjugation the ultimate goal of modern culture and its inevitable fate. Yet even if we think it impossible to constrain or stop this course of things, a final question remains. It belongs to the essence and determination of mind[2] not to tolerate any external determination. Even where it entrusts itself to a foreign power and sees its progress determined by it, the mind must at least attempt to penetrate the core and meaning of this determination. Thereby mind reconciles itself with its fate and becomes free. Even if the mind is not able to repel and conquer the power to which it is subjected, it nevertheless demands to know this power and to see it for what it is. If this demand is made in earnest, it does not possess a purely 'ideal' significance and is not limited to the realm of 'pure thought'. From the clarity and certainty of seeing follows a new strength, a power or efficacy, a strength with which mind strikes back against every external determination, against the mere fatality of matter and the effects of things. Insofar as mind considers the powers that seem to determine it externally, this consideration already contains a characteristic turning back and turning inward. Instead of grasping outwardly at the world of things, it now turns back onto itself. Instead of exploring the depths of effects, it returns to itself and, by means of this concentration, achieves a new strength and depth.

Admittedly, we are today still far away from fulfilling this ideal demand, particularly in the realm of technology. A gulf repeatedly emerges that separates thinking from doing and knowledge from action. If Hegel is correct when he states that the philosophy of an age is nothing more than that very age 'grasped in thought', and if this philosophy, understood as the concept of the world, only appears after reality has completed this process of formation and so 'finished itself',[3] then we would have to expect that the incomparable development which technology has undergone over the course of the last century corresponds to a change in the way we think. However, if we look at philosophy's present situation, this expectation has been only incompletely fulfilled. Admittedly, from approximately the middle of the nineteenth century onwards, problems which had their origins in the area of technology have increasingly made their way into abstract 'philosophical' examinations, thereby giving them a new goal and direction. Neither the philosophy of science nor value theory has escaped this influence. The theory of knowledge, the philosophy of culture and metaphysics all attest to technology's breadth and growing power. This relation presents itself most clearly in certain currents of the modern theory of knowledge, which attempt to transform the traditional relationship between 'theory' and 'praxis' into its opposite, defining theoretical 'truth' merely as a special case of 'utility'. Beyond these properly 'pragmatic' trains of thought, the growing influence of technological concepts and questions on philosophy as a whole is unmistakable. Even modern Lebensphilosophie is often subject to it, though Lebensphilosophie believes it takes the most vigorous stand against it. It too is not free from the chains it mocks. But all of these inevitable points of contact between the realms of technology and philosophy in no way prove that an inner communality is being initiated and built up between the two. Such a community can never result from a mere sum of external 'influences', however manifold and strong we may think them. That philosophy and technology have jointly entered into the systems of positivism and empiricism – we need only think of Mach's principle of economy as the basis of a theory of knowledge – should not be taken as a certification proclaiming a true unification of the two. Such a unification would be reached only if philosophy succeeded fulfilling on this point the general function that it has increasingly fulfilled with ever-greater clarity for the other spheres of culture. Since the days of the Renaissance, philosophy has brought all the powers of modern thought before its forum, questioning them about their meaning and right, their origin and validity. This question of validity, of the quid juris as Kant calls it, is directed to all the formal principles of thought; in posing this question, the grounds of their specific characteristics first become uncovered, their own proper meaning and value discovered and assured. Philosophy has achieved such assurance, such 'critical' consciousness and justification, for mathematics, the theoretical knowledge of nature, the 'historical' world and the humanities. Although new problems constantly arise here, although the work of 'critique' shall never come to an end, the direction of this work has been set since the days of Kant and his founding of 'transcendental philosophy'. Technology, however, has not yet seriously been integrated within this circle of philosophical self-reflection. Technology still seems to retain a singularly peripheral character. Even though technology has expanded beyond the periphery, genuine knowledge of technology, insight into its 'essence', has not kept pace. A fundamental motive for the inner tension and antagonism found in the formative tendencies of our epoch lies precisely in this disparity: 'abstract' thought is unable to penetrate into the core of the technological world. A resolution of this tension can never be hoped for or sought by adjusting the extreme points of the tension or effecting a mere compromise between them. Rather, a possible unity requires acknowledging that this particular case involves more than a mere difference. It is a genuine polarity. This fact determines the task that philosophy has to fulfil with respect to the current development of technology. The task cannot be limited to assigning technology a predetermined 'place' in the whole of culture and, therefore, in systematic philosophy that aims to be the intellectual expression of culture. Technology cannot simply be placed next to the other areas and entities, such as 'economics' and 'the state', 'morality' and 'law', 'art' and 'religion'. In the realm of culture, separate areas never stand simply together or next to one another. Here, the community is never spatially static but possesses a dynamic character. One element is found 'with' the other only to the extent that both assert themselves in opposition to each other and thereby mutually confront and determine each other. Thus, every introduction of a new element not only widens the scope of the mental horizon in which this confrontation takes place, but it alters the very mode of seeing. This formative process does not only expand outwardly – it itself undergoes an intensification and heightening, so that a simultaneous qualitative transformation occurs, a specific metamorphosis. It is not enough for modern philosophy simply to find a 'space' for technology in the edifice of its doctrine. A space that is created in this way will always remain an aggregate space and never become a truly systematic one. If philosophy wants to remain loyal to its mission, if it wants to maintain its privilege, so to speak, of representing the logical conscience of culture, it must also enquire into the 'conditions of the possibility' of technological efficacy and technological formation, just as it enquires into the 'conditions of the possibility' of theoretical knowledge, language and art. Here too, philosophy will be able to ask the question of being and the question of validity only when it has clarified the question of meaning. However, this clarification cannot succeed so long as one's observations are limited to the circle of technological works, to the region of the effected and created. The world of technology remains mute as long as philosophers look at it and investigate it from this single point of view. It begins to open up and to divulge its secret only if we return from the forma formata to the forma formans, from that which has become to the very principle of becoming.

Today the need to return to this principle is felt much more by those who work in technological fields and are engaged in its productive labour than by those who work in systematic philosophy. In technology the power of 'materialistic' ways of thinking and questioning has been given up. The search for the purpose and legitimacy of technology requires posing this question ever more clearly and ever more consciously in reference to the 'idea' it embodies. 'The origin of technology', as expressed in one of the newest works in the philosophy of technology, 'lies in the idea'.[4] To cite another author: 'We will look at technology as the organic partial appearance of a larger phenomenon, the development of culture itself. We will attempt to understand it as the embodied expression, as the historical fulfillment of a basic idea required for a system of cultural ideas where the tangible material of technological creations comes to be inwardly mastered – regardless of how varied the expression of the idea is in the battle of motives and tendencies among those engaged in these activities. The task is to recognize the *transpersonal as an ideal unity* or joint effect that determines human actions – not as a kind of blind law, but as something they freely take up, in order to ... become historically effective.'[5] Whatever the answer, the question itself is thereby transferred to the level where all genuine mental decisions belong. The question also leads the problem back to its initial historical origin and is linked to it in a remarkable and surprising way. Just as a modern thinker standing in the midst of the concrete technological forms of life comes to see the crux of the problem, so too the discoverer of the 'idea' and the 'world of ideas' conceived it over 2,000 years ago. When Plato develops the relationship between 'idea' and 'appearance' and seeks to justify it systematically, he does not seek to ground it in the shapes of nature but in the products and organization of τέχνη.[6] The art of the 'craftsman', the 'demiurge', provides him with one of the great motifs with which he represents the meaning of the idea. According to Plato, this art is no mere imitation of something that is already simply present. This art is possible only on the basis of a prototype and archetype to which the artist looks in his creative work. The artist who first invented the loom did not initially find it as something given in the sensible world; rather, he introduced it into the sensible world by looking towards its form and purpose, to its eidos and telos. Today, the constructor of the loom still looks to the form. For instance, if a loom is broken and a new one must be constructed, the broken loom is not used as a model and pattern; rather, what gives direction to the constructor's new work is his gaze upon the original form as exhibited in the mind of the first inventors. Thus, this general form, not an individual thing existing in the sensible world, constitutes the actual 'being' of the loom.[7] Is it a coincidence, then, that this basic tenet of Platonism is also increasingly asserting itself in contemporary reflections on the meaning of technology? Dessauer, for example, remarks that, 'from a higher sphere of reality and power, through the mind and hands of the technician and worker, an immense stream of experience and power descends into earthly existence. A spiritual stream pours into the chaotic material world, and everyone, from the creator to the final worker, takes part: all are recipients.' Similarly, Max Eyth argues that, '*Technology is every-thing which gives the human will an embodied form.* Here, human willing coincides with the human mind, which contains an unending number of life-externalizations and life-possibilities. Technology, despite being bound to the material world, also received something of the boundlessness of the pure life of mind.'[8] Such remarks clearly illustrate that modern attempts to make sense of the basis and essence of technology are no longer satisfied to view it merely as an 'applied natural science' which is somehow harnessed and captured in the concepts and categories of natural science. What is sought, rather, is technology's relation to cultural life in its totality and universality. This relation, however, is to be found only when we focus on the concept of form rather than the concept of being as understood in the natural sciences, and when we reflect on the ground and origin of the concept of form, its content and meaning. The concept of form first opens the expanses of thought to us and determines the horizons of the mind for us.[9] If, instead of beginning from the existence of technological objects, we were to begin from technological efficacy and shift our gaze from the mere product to the mode and type of production and to the lawfulness revealed in it, then technology would lose the narrow, limited and fragmentary character that otherwise seems to adhere to it. Technology adapts itself – not directly in its end result but with a view to its task and problematic – into a comprehensive circle of enquiry within which its specific import and particular mental tendency can be determined.

In order to penetrate this circle and truly grasp its core, another fundamental and purely methodological reflection is needed. The particular character of the question of meaning that confronts us here repeatedly threatens to become obscure; its borders repeatedly threaten to become blurred because of other motives that not only join it but also gradually and imperceptible lead to its displacement. Such a displacement has already occurred if we believe that the question of meaning can be equated with the question of value – and that such a starting point can bring about a genuine solution to the question. In this identification of 'meaning' and 'value', a deferral of the problem has already taken place. Admittedly, this logical lacuna not only goes unnoticed inasmuch as it is found in connection to the problem being investigated here, it also pervades the whole expanse of the 'philosophy of culture' and spans the totality of its tasks. So often in the history of thought, the 'transcendental' question is posed about the 'possibility' of culture, its conditions and principles; but rarely has this question been held onto and explored with great acuity, especially concerning its pure essence. It constantly flits away in two different directions: the question concerning cultural achievement has been subordinated to the question concerning its content. While we might like to measure this achievement according to different mental dimensions, this would not rectify the mistake already committed in the first formulation of the problem, no matter how high or how low we might estimate it. This state of affairs already emerges with the first real 'critic' of modern culture, Rousseau. When Rousseau placed the intellectual culture of his time before the real questions of conscience and destiny, the framing of his question was dictated by external sources, the competition sponsored by the Academy of Dijon in 1750. The question was whether the rebirth of the Arts and Sciences had contributed to the ethical perfection of humanity (*'Si le rétablissement des sciences et des arts a contribué à épurer les mœurs').*[10] In the mind of Rousseau, which was in accord with the basic orientation of Enlightenment ethics, this perfection was reached by fulfilling desire and enjoying a standard of 'happiness' won through humankind's transition from the state of 'nature' to that of culture. 'Happiness' and 'perfection' are the two dimensions within which he seeks the answer to his problem. They provide the standards by which his responses are to be adjudicated. It was not until German Idealism that a crucial turn was brought about; German Idealism was the first to pose the 'question of essence' with great acuity and clarity, disengaging it from the additional questions of happiness and moral 'perfection'. Thus, for instance, in the *Critique of Judgment* the realm of the beautiful could be philosophically justified through the autonomy – the self-legislation and self-signification – of the beautiful, which is discovered and guaranteed in opposition to the feelings of pleasure and displeasure as well as the norms and rules of the ethical 'ought'. If we turn to the realm of technology and to the ever-intensifying struggle that goes on within it in order to understand its specific meaning and content, we discover that the struggle remains for the most part at a preliminary stage, a stage the o ther areas of culture have long since passed through. We may bless technology or curse it, we may admire it as one of the greatest possessions of the age or lament its necessity and depravity – in judgements such as these, a measure is applied to it that does not originate from it. Consciously or unconsciously, purposes are ascribed to it that are foreign to technology's pure formative will and power. And yet an authentic judgement can come only from within technology itself, that is, only from insight into its own inherent, immanent law. The p hilosophy of technology, at least, is tied to this demand. Admittedly, philosophy also confronts the contents of culture not only by observing and testing them but also by judging them. It does not want to merely know them, but also to acknowledge and dismiss, judge and assess, decide upon and direct them. This philosophy can and must do. Its intellectual conscience, however, forbids it to make a judgement before it has penetrated into the essence of that which is being judged, grasping it on its own terms. This freedom of the philosophical gaze, however, can hardly ever be found in modern apologies for technology and in the attacks and accusations that are directed against it. Again we are tempted to employ the maxim that S pin o za formulated in his political philosophy for both the accused as well as the plaintiff: '*[N]on ridere, non lugere, neque detestari; sed intelligere.*'[11] The determination of 'being' and 'being-such-and-such', the consideration of what technology is, must precede the judgement of its value. Here arises a new dilemma: the 'being' of technology permits itself to be grasped and represented in no other way than in its activity. It appears only in its function. It consists neither in its external appearance nor in what it externalizes; rather, it consists in the manner and direction of the externalization itself, in the formative impulse and process, which this externalization is subject to. Thus, being can become visible only in becoming, work can become visible only in energy – and this particular difficulty clears the way and indicates the direction for further consideration. Exactly here at this point, the affinity and internal connections that exist between technology and the pure form and principle of other basic powers of culture become clear, no matter how different they may be with respect to their content. What Humboldt has said and proven for language is also valid for these other powers: the genuine conceptual determination, the only true 'definition' that can be given for these powers, is a genetic one. They can and must not be understood as a 'dead product' but as a way and basic direction of production. It is from within this intellectual perspective that we should enquire into the essence of technology. Goethe says that when a human being acts meaningfully, he always and simultaneously acts as a law-maker. It belongs to the essential task of philosophy to penetrate into this human law-giving, to gauge its unity and internal differences, its universality and differentiation. Only through such a comprehensive endeavour can we obtain a secure basis for a detailed judgement; only then can we hope to obtain a norm raised above all merely subjective expressions of praise and reprimand, favour and displeasure, seizing instead the genuinely objective 'form' of the perceived object in its nature and in its necessity.

II

Max Eyth, one of the most enthusiastic and eloquent pioneers of the cultural autonomy of technology, begins his lecture 'Poetry and Technology' from the known kinship between the function of technology and the function of language:

> Two things essentially distinguish animals from human beings, understood from the perspective of their external appearance: the word and the tool. The ability to create words and tools has ... made the human being out of the animal. How these abilities have come into the world will undoubtedly remain an eternal puzzle that no theory of evolution will be capable of solving, because they originate in a wellspring from which no animal ... has ever drunk. Both abilities were imperative for the survival of the human being in a hostile

world in which he, physically more helpless, weaker, and less resistant than most animals, would undoubtedly have quickly perished. What saved him ... in the sphere of knowledge was language; in the sphere of ability, the tool ... The power that turned the mere defenseless human being into the sovereign over every living thing on Earth rests on knowledge and ability, on the word and the tool ... In prehistoric times, far from the beginnings of culture, the tool undoubtedly played the primary role in the formation of human existence ... Later ... a decisive alteration in the relationship between word and tool emerged. Language, just because it can speak, knew how to create for itself an outstanding, one could even say unwarranted, significance. For mankind, mute tools were increasingly relegated to the background. Knowledge was master and ability served. This relation continued to intensify and has continued to be accepted until now. Today we stand amid a fierce struggle that is endeavoring, if not to alter, then to return the relation of the two to its proper foundation. In its growing domination, language ... exalted its unwarranted claim to be the only 'tool of the mind'. ... In general, language believes this still today. Concerning the 'tool of the mind', language forgets the mental aspect of the physical tool. Both word and tool are a product of the same fundamental mental force that has made the animal 'homo' into the human being, 'homo sapiens', as it is called by the scholars who, of course, allude only to the human being's knowledge and forget the skill that has rendered all his knowledge possible.[12]

I have singled out these sentences by a technician and a thinker of technology because a real philosophical problem is hidden in the parallel asserted here between language and tools. It is not merely wit, or an external analogy, that brings together language and tools and attempts to understand them by one principle. The idea of such an essential relation was not foreign to the first 'philosophers of language' within the sphere of our European thought. They did not believe that words and language were primarily means of representation, means for the description of external reality. Rather, they saw in language a means for the making of reality. For them, language became a weapon and tool human beings employed in order to compete in the struggle with nature and with their peers in social and political conflict.[13] 'Logos' itself, as the expression of the particular mental nature of the human being, appears here to have an 'instrumental' as well as a 'theoretical' meaning. Yet implicitly contained in this is the counter-thesis that the potency of logos also resides in every simple material tool, in every application of a material thing that serves the human will. Thus, the determination of human essence, the definition of the human being, develops in this twofold direction. The human being is a 'rational' being in the sense that 'reason' comes from language and is insolubly bound to it; ratio and oratio, speaking and thinking, become interchangeable concepts. At the same time, and no less originally, man appears as a technological, a tool-forming being: 'a tool-making animal',[14] to employ Benjamin Franklin's words. The power with which man asserts himself against external reality, and by virtue of which he first gains an intellectual image of this reality, is determined by these two sides of his essence. All mental handling of reality is bound to this double act of 'grasping' – 'comprehending' reality in linguistic-theoretical thought and 'gripping onto it' through the medium of efficacy. This is true for both mental and technological forming.

In both cases it is essential to guard against a misunderstanding in order to penetrate into the actual sense of this forming. The 'form' of the world, whether in thought or action, whether in language or in effective activity, is not simply received and accepted by the human being; rather, it must be 'built' by him. In this respect, thinking and doing are originally united, they both stem from this common root of forming gestalts, gradually unfolding and branching off from it. Wilhelm von Humboldt[15] has shown this basic relationship in language. He demonstrates how the act of speaking is never a mere receiving of the object, a reception of the existing form of the object in the I. Rather, it contains in itself a real act of world-creation, the raising-up of the world to form. The notion that different languages only denote the same mass, independent of the objects and concepts available to them, is, for Humboldt, truly pernicious for the study of language. This view masks that which constitutes language's genuine meaning and values. It conceals language's creative role in the laying out, production and securing of the concrete view of the world. The difference among languages is not a difference between sounds and signs. Rather, it is 'a difference of world views'.[16] Correctly understood, what is said here about the use of language also holds for each use of the material tool, however elementary and 'primitive'. Here, too, that which is crucial is never found in the material goods that are gained through it, in the quantitative expansion of the sphere of influence through which, little by little, one part of external reality after another is submitted to the will of the human being. The will that initially seemed limited by its proximity to the human body, to the movement of its own limbs, gradually explodes and breaks through all spatial and temporal barriers. In the end, this overcoming would be fruitless if it contained and dragged along with it only new world-matter. Here, a more genuine and greater profit lies in the gaining of 'form', in the fact that the expansion of efficacy brings about a change in its qualitative meaning, creating the possibility of a new aspect of the world. Efficacy, in its continuous increase, in its expansion and intensification, would finally have to be recognized as powerless, as internally aimless and weak, if an inner transformation, an ideal turn in its meaning, were not simultaneously being prepared and constantly carried out. What philosophy is able to achieve for technology, for its understanding and legitimacy in thought, is the demonstration of this turn in meaning. To do this, philosophy must grasp deep into the past. It must seek to penetrate back to when the secret of the 'form' first opens itself to the human being, when it begins to rise up in thought and deed – in order, admittedly, to cloak itself just as much as to reveal itself – so as to exhibit itself only as in a puzzling mist, in the 'twilight of the idols' of the magical-mythical worldview.

If we compare the worldview of various so-called civilized cultures to indigenous tribes, the deep opposition that exists between them reveals itself perhaps no more sharply than in the direction the human will adopts in order to become master over nature and gradually to take possession of it. A type of magical desire and efficacy confronts technological will and accomplishment. People have sought to derive this original opposition from the totality of differences that exists between the world of civilized people and indigenous people. Humans from an earlier time are distinguished from those of a later time, just as magic is distinguished from technology. The former may be denoted as *homo divinans* and the latter as *homo faber*. The whole development of humanity presents itself, then, as a completed process containing innumerable intermediary forms, through which the human being moves from the initial stage of *homo divinans* to the stage of *homo faber*. If we accept this distinction that Danzel has forcefully maintained and carried through in *Kultur und Religion des primitiven Menschen*,[17] we haven't reached a solution to the problem. We have only formulated it. For it would only be an assertion and extrapolation if ethnology, from which this distinction originates, attempts to explain it by attributing to 'magical' man a predominance of 'subjective' determinations and motives more than purely 'objective' ones. The worldview of *homo divinans* is supposed to come about through the projection of his condition onto reality; he sees in the external world what is going on within himself. Inner processes that take place entirely within the soul are transferred outside of the human body. Drives and wilful movements are interpreted as strengths that intervene directly into events, steering and

altering them. However, from a purely logical perspective this explanation is marred by a petitio principii – it confuses that which is
to be explained with the ground of explanation. When we reproach
indigenous peoples for 'confusing' the objective and subjective, for
letting the borders of both areas flow into one another, we are speaking from the standpoint of our theoretical observation of the world
founded on the principle of 'cause', on the category of causality as the
condition of experience and the objects of experience. These borders
are not 'in themselves' objectively before us; rather, they must first be
set down and secured, they must first be erected by mental labour.
The manner of setting these borders takes place differently according to
the position in which mind finds itself and according to the direction
in which it moves. Every transition from one posture and direction to
another always ends in a new 'orientation', a new proportion between
the 'I' and 'reality'. Thus, the relation between both is not set down
as unique and unambiguous from the beginning. It first comes to be
because of the manifold ideal processes of 'mutual differentiation and
determination', as in myth and religion, language and art, science and
the different basic forms of 'theoretical' conduct in general. For human
beings, a fixed relation of subject and object according to which they
conduct themselves does not exist from the beginning. Rather, in the
entirety of a human being's activity, in the entirety of his bodily and his
psycho-spiritual activities, there first arises knowledge of both subject
and object; the horizon of the 'I' first separates itself from that of reality.[18] There is no solid, static relation between them from the outset.
There is, as it were, a fluctuating movement of back and forth. From this
movement a form gradually crystallizes in which the human being first
grasps his own being as well as the being of objects.

    If we apply this general insight to the problem that is present here,
we see that the human being, in his magical and technological activity,
does not already have a determined form of the world. He must instead
search for this form and find it in various ways. The way he finds it
depends on the dynamic principle that the general movement of mind
follows. If we assume that the principle of 'causality' and the question
concerning the 'ground' of being and the 'causes' of events already
prevail in the magical view of nature, the partition between magic and
science falls away. In his work *The Magic Art*, James George Frazer,
one of the best specialists on magical phenomena, expressly draws this
conclusion in his attempt to lay out completely the factual sphere of
the magical arts. At the same time, he links a certain theory about the
meaning and origin of magic to his description of this factual sphere.

On Frazer's account, magic amounts to nothing other and nothing less than the beginnings of 'experimental physics'. In magic, the human first perceives objective being and happening, which are ordered according to fixed rules. The course of things now appears to him as a closed nexus, a chain of 'causes' and 'effects' in which no supernatural power can arbitrarily intervene. According to Frazer, it is here that the world of magic is clearly separated from the religious world. In the religious outlook, the human is subjected to a foreign power to which he entrusts the whole of his being. Here there is still no fixed natural course, for the world still does not have its own gestalt and its own power; it is a plaything in the hands of superior transcendent powers. It is, however, just this basic view against which the magic worldview protests. It grasps nature as a strictly determined sequence of events and seeks to penetrate into the essence of this determination. It knows no coincidence. It rises to the conception of a strict uniformity of events. And, in this way, it achieves, in contrast to religion, the first stage of scientific knowledge of the world. Magic admittedly differs from science in its result but not in its principle and its problem. This is the case because the principle 'like causes, like effects' governs it as well, giving it its generally apparent character. That it is not able to employ this principle in the same sense as the theoretical science of nature is not, according to Frazer, due to a logical reason but only to a factual one. It is 'primitive' not in its form of thought but in the measure and the security of its knowledge content. The circle of observation is too narrow, the nature of observation too fluctuating and uncertain, for it to be able to erect truly durable empirical laws. The consciousness, however, of lawfulness as such has been awakened in it and is tightly and steadfastly held onto by it. Thus in the end, Frazer sees in both basic forms of magic nothing other than the applications and variation of the 'scientific' principle of causality, which he understands and expounds here in accordance with the views of English empiricism: 'sympathetic' magic and 'homoeopathic' or 'imitative' magic are both founded on the fundamental laws of ideal association that rule over all causal thinking. In the case of the former it results in the law of 'association by similarity' and in the case of the latter it results in the law of 'association by contact' and becomes the guiding principle of theoretical and practical activity.[19]

The flaw in Frazer's theory, which is endorsed by a great number of ethological researchers, can be stated as follows: it awards magical activity a significance and ascribes to it an achievement that is reserved for technological activity. Magic may differ from religion insofar as the human being is able to escape the merely passive relationship to nature – that is, he no longer receives the world as the mere gift of a superior divine power but wants to take possession of it and stamp it with a determined form. But the manner of this appropriation is entirely different from the appropriation carried out by technological efficacy and in scientific thinking. The magical human being, the *homo divinans*, believes in a certain sense in the omnipotence of the 'I'. However, this omnipotence expresses itself only in the force of a wish. Reality is not able to withdraw from wishing in its highest intensification and potency; it is connected and subjected to it. The success of a particular act is linked to reality in the following way: the goal of the action is precisely anticipated in the imagination, and the resulting image of this goal is worked on and held to with great intensity. All 'real' actions, if they are to be successful, need such a magical preparation and anticipation. Warring or raiding, fishing or hunting, can succeed only if every phase is magically anticipated in the right way and at the same time 'rehearsed'.[20] Already in the magical worldview, the human being tears himself away from the immediate presence of things and builds his own kingdom with which he reaches out into the future. However, if in a certain sense he is freed from the power of immediate sensation, then he has only exchanged it for the immediacy of desire. In this immediacy, he believes he is able to seize reality directly and to conquer it. The totality of magical practices is, so to speak, simply the laying out, the progressive unfolding of the desired image that the mind carries within itself of the goal to be reached. The simple, ever more intense repetition of this goal is already regarded as the way that must inevitably lead to it. Herein originate the two archetypes of magic: word-magic and image-magic. Word and image, then, are the two ways in which the human being handles a non-present thing as present – by which he, as it were, sets something wished and longed for before himself, in order, in this very act of 'imagination', to enjoy and to make it his own. That which is spatially remote and temporally distant is 'called forth' in speech or is 'imagined' and 'prefigured'. Already here, the *regnum hominis*[21] is sought-after, though it slips away at once and dissolves into a mere idol. Undoubtedly, magic is not merely a way of world-apprehension, but contains within it real seeds of world-formation. But the medium in which it moves does not let these seeds develop, for the reality of experience is still not seen in its order and rules. It is enveloped more densely into a simple, wishful dream that conceals its own form. Moreover, this accomplishment of 'subjectivity' is not to be assessed in an exclusively negative fashion, for it is already a first and, in a certain sense, a crucial step. The human being does not simply abandon and submit himself to an impression of things, to their mere 'givenness', but changes them, letting a world be generated out of himself. When he is no longer satisfied by mere existence, he demands to be a something and to be different. However, this first active direction in which the world of being faces the world of doing still lacks the means of actuation. Because the will jumps directly towards its goal in the magical identification of 'I' and 'world', no true mutual determination between them occurs. For every such confrontation calls for proximity as well as distance, empowerment as well as relinquishment, the force of grasping but also the force of keeping something remote.

It is precisely this double process revealed in t e c h n o l o g i c a l a c t i v-i t y that differentiates it from magical activity. Here, the power of the will replaces the power of mere desire. This will reveals itself not only in the force of the forward-driving impulse but also in the way in which this impulse is led and mastered. It reveals itself not only in the ability to seize its goal but also in the particular ability to distance the goal from it and to leave it at this distance, letting it stand there. It is only this letting-stand of the goal that makes an 'objective' sense perception possible, a sense perception of the world as a world of 'objects'. For the will, the object is just as much the guiding principle and thread that first gives it its determination and its solidity, as it is the limit of the will, its counterpart and its resistance. The strength of the will first grows and becomes stronger on the strength of its limit. The will can never succeed in its application simply by making itself stronger. Success demands that the will intervene in an originally foreign order and that it know and recognize this order as such. This knowing is at the same time a mode of recognition. Nature is not, as in magic, merely repressed by desiring and imagining. Rather, its own independent being is acknowledged. And the true victory of thought is only achieved in this self-modesty. '*Natura ... non nisi parendo vincitur*':[22] victory over nature is only achieved through obedience to it. By means of this obedience, which lets nature prevail and no longer seeks to captivate and subjugate it magically, a new gestalt – in a purely 'theoretical' sense – of the world emerges. Human beings no longer attempt to make reality amenable to their desires with various methods of magic and enchantment. They take it as an independent and characteristic 'structure'. In this way, nature has ceased to be an amorphous material that yields to every metamorphosis and, in the end, allows itself to be forced into any gestalt through the power of magical words and images. In place of magical compulsion, the 'discovery' of nature emerges, which is contained in all technological activity, no matter how simple and primitive the application of the tool may be.

This discovery is a disclosure; it is the grasping and the making one's own of an essential connection that previously lay hidden. Thus only here are the fullness and limitless changes of the gestalts of the magical-mythical world traced back to a determined, standard measure. Yet, on the other hand, reality does not become rigid through a reduction to its inner relation of measure; its inner mobility has been preserved and has lost nothing of its 'plasticity'. However, this plasticity, this 'form-ability', is now set as if in a fixed intellectual framework limited by certain rules of the 'possible'. This objective possibility now appears as the border where the omnipotence of desire and affective fantasy are placed. In place of merely libidinous desire, there first emerges a genuine, conscious wilful relationship – a relationship in which ruling and serving, demanding and obeying, victory and submission are united. In such a mutual determination, a new meaning of the 'I' and a new meaning of the world are grasped. The arbitrariness, self-will and obstinacy of the I withdraw, and insofar as this happens the proper meaning of Dasein and happening, reality as cosmos – as order and form – stand out.

To make this clear, we need not look at the complete unfolding and present structure of technology. A basic circumstance presents itself in the most ordinary and inconspicuous phenomena, in the first and simplest beginnings of tool-use, more clearly than in almost all the marvels of modern technology. Already here we penetrate, from a purely philosophical perspective, into the core of the problem. Although the distance between the most cumbersome and imperfect tools we use and the results and achievements of technological execution appears vast, at least with respect to their content, if we focus on the principle of action, we find that the gap is much smaller than the gulf that separates the first invention and application of the crudest tool from mere animal behaviour. It would not be an exaggeration to say that the transition to the first tool not only contains the seeds of a new mastery of the world, but also a turning point in knowledge. The mode of action established here grounds and steadies, for the first time, a type of mediacy that belongs to the essence of thought. In its pure logical form, all thought is mediated. It is directed to the discovery and extraction of a mediating structure, which joins the opening sentence and the ending sentence of a communicative chain. The tool fulfils the same function, represented here in the logical sphere, in the objective sphere of physical objects. It is grasped, as it were, in objective sense perception; it is not merely the *'terminus medius'*[23] of thinking. It sets itself between the first positions taken by the will and its goal. Only in this in-between position is it permitted to separate them and set them at a proper distance. As long as the human being makes use only of his limbs, his bodily 'organs', in order to achieve his goals, such distancing is not yet reached. Admittedly, he effectively acts on his environment. But there is a great distance between this work and the knowledge of this very efficacy. Whereas all human doing is absorbed in apprehending the world, human beings cannot yet comprehend it as such, because they cannot yet conceive of it as an objective gestalt, as a world of objects. The elementary taking-possession-of, immediate physical seizing, is not a constructive grasping. It does not lead to a building up in the region of sense perception or in the region of thought. In the tool and its application, however, the goal sought-after is for the first time moved off into the distance. Instead of looking spellbound at this goal, the human being learns to 'fore-see' it. This 'fore-seeing' becomes both means and condition for attaining the goal. This form of seeing is all that distinguishes human intentional doing from animal instinct. This 'fore-seeing' establishes 'fore-thought'. It establishes the possibility of directing attention to a goal, towards something spatially absent and temporally remote, rather than acting on an immediately given sensuous stimulus. It is not so much because animals are inferior to the human in bodily skill. But because this line of sight is denied to animals, there is no genuine tool use in the area of animal existence.[24] And it is also from this line of sight that there first arises the thought of causal connection in the strict sense of the word. If one takes the concept of causality so loosely that it can be present wherever spatial and temporal co-extension connects through mere 'association', then the origin of this concept must be considered to be much earlier. There is no doubt that association is present in the magical act and that the magical world is pervaded by it. Frazer follows this view of causality when he subordinates the world of magic to the principle of causality, when he sees in magic the true beginning of 'experimental physics'.[25] But another picture – and judgement – of the logical connections and differences between the basic forms of world understanding emerges if we take the concept of causality in the sharper and stricter sense Kant gave to it in his criticism of Hume's theory of causality. The main focus of this critique lies in the proof that it is in no way the mere 'habitual' connection but the thought of a 'necessary' connection that determines the nucleus of the concept of causality as a category of the 'pure understanding'. And the correctness of this notion is to be sought-after and proven by showing that, without it, the relation of our ideas to an object would not be possible. The concept of causality belongs to the original forms of synthesis, which alone make it possible to give ideas an object. It is, as the condition of the possibility of experience, the condition of possibility of the objects of experience. The mythical-magical world still knows nothing about a sense of causality that both constructs and renders possible the sphere of objects, making them accessible to thought. For the mythical-magical world, the whole of nature is similarly broken into a play of forces, into actions and reactions. These forces, however, are of the sort that the human being lives with and experiences in his immediate drives. They are personal, demonic-divine powers that direct and determine events, and whose participation human beings must secure in order to influence these events. With the creation of the tool and by means of its regular use, the limits of this type of representation were first breached. Here we encounter the 'twilight of the gods' of the magical-mythical world. Only here does the notion of causality emerge from the limitations of 'inner experience', from being bound to the subjective feelings of the will. It becomes a bond that joins pure objective determinations together and sets down a fixed rule for their mutual dependence. The tool no longer belongs immediately, like the body and its limbs, to the human being. The tool signifies something detached from its immediate being and becomes something that exists in itself, a continued existence that can far outlast the life of the individual human being. This kind of 'thing-hood', this 'reality', does not, however, now stand alone; it is truly real only in and through the effects it wields on other beings. These beings are not simply joined externally to the tool. They belong to its particular essence. The perception of a particular tool, for instance the perception of an axe or a hammer, never exhausts itself in the perception of a thing with particular characteristics, of materials with certain qualities. Here, its use – its function – becomes apparent in its very stuff. The form of its activity comes to be in 'matter'. They are not separated from one another but are apprehended and comprehended as an insoluble unity. The object is determined as something only insofar as it is for something. This is because in the world of tools there are no mere things with properties. There are only ensembles of 'vector-magnitudes', to use a mathematical expression. Although every being is determined here in-itself, it is, at the same time, the expression of a particular activity to be performed. And in the perception of this activity, a fundamentally new direction of seeing opens up for the human being: the perception of 'objective causality'.

Of course, when we consider this achievement, we should bear in mind that the gap between the two different aspects of the world confronting one another cannot be jumped over all at once. The distance between the two poles continues to exist and can be traversed only step by step. Long after the human mind has produced, in both language and tools, the most important means of its liberation, these methods still appear enveloped in the magical-mythical atmosphere which it is supposed to overcome in its final and highest development.[26] The world of language, like that of tools, is in no way immediately comprehended as the creation of the human mind, but rather as the efficacy of foreign and superior forces. The demonic character that belongs to the mythical conception as such also includes these two worlds and at first threatens to draw them completely under its spell. The totality of language and tools appear as a kind of pandemonium. Originally, language is not the means of a matter-of-fact presentation, a medium for the exchange of information that serves to bring about reciprocal, logical understanding. The more we attempt to return to the 'origins' of language, the more its purely 'objective' character is lost. Herder says that the oldest dictionary and grammar of humanity were nothing more than a 'pantheon of tones', a realm consisting less of things and their names than of animate, acting beings. The same held for the first and most primitive tools. They too are regarded as 'given from above' as gifts from a god or saviour. They are worshipped as divine. The Eweer tribe in South Togo still regards the blacksmith's hammer as a mighty deity, to which they pray and offer sacrifice. The traces of this feeling can be seen in the great cultural religions.[27] But this awe subsides. The mythical darkness that still surrounds the tool gradually begins to clear to the degree that they are not only used but also, through this very use, continually transformed. So the human becomes increasingly conscious of being a free sovereign in the realm of tools. Through the power of the tool the tool-users come, at the same time, to view themselves differently, now as the administrator and producer of the tool. 'The human being experiences and enjoys nothing', says Goethe, 'without at the same time being productive. This is the innermost quality of human nature. We can even say without exaggeration that it is human nature itself.'[28] This basic force of the human being reveals itself perhaps nowhere as clearly as in the sphere of the tool. The human works with it only insofar as he, in some way, even if initially with only modest results, works on it. It is not merely his means for transforming the objective world – in the process of the objective world's metamorphosis the tool itself undergoes a transformation and moves from place to place. And in this change the human now experiences a progressive increase, a peculiar strengthening of his self-consciousness. A new world-attitude and a new world-mood now announce themselves over and against the mythical-religious worldview. The human being now stands at that great turning point in his destiny and self-knowledge that Greek myth embodied in Prometheus. Titanic pride and consciousness of freedom confront fear and reverence for demons and gods. The divine fire is wrested away from the seat of the immortals and placed in the sphere of the human being, in his home and hearth. The world of desire and dreams in which magic had enveloped the human being is destroyed. Man sees himself led into a new reality that receives him with a seriousness, and severity, and necessity that obliterates all his desires. However, if he cannot escape this necessity, and he is no longer able to control the world according to his desires, he now learns to master it increasingly with his will. He no longer attempts to control its course; he falls into line with the iron law of nature. But this law does not enclose him like the walls of a prison. By means of this law, he tests and wins a new freedom. For reality shows itself, regardless of its strict and irrevocable order, not as an essentially rigid existence but rather as a modifiable, malleable material. Its gestalt is not complete. Rather, it offers human will and initiative enormous latitude for action. And it is by moving about in this space, in the whole of that which is achieved through his work – and through which his work first becomes possible – that the human progressively builds up his world, his horizon of 'objects', and the concept of his own essence. He now sees himself expelled from that magical realm of immediate wish-fulfilment that magic has enticingly placed before him. He is expelled onto a limitless path of creative work that promises him no essential goal, no more final stop or resting point. However, in lieu of all this, a new determination of value and meaning is now established for his consciousness: the genuine 'purpose' of action is no longer measured by what it brings about and finally achieves; rather, it is the pure form of doing, the type and direction of the productive force as such, that determines this purpose.

III

The indispensable participation of technological creation with the conquest, securing and consolidation of the world of 'objective' sense perception has become clearer through the preceding observations. It has become increasingly clear that a certain misgiving not only threatens to problematize the value of technological achievements but also to turn them directly into their opposite. Is not what was regarded here as the authentic achievement of technology nothing other than the basic evil from which it suffers? Does not this increased accessibility to the world of objects at the same time necessarily result in the alienation of human beings from their own essence, from what they originally are and what they originally feel? With the first step into the world of facts that technological work secures and constructs for him, the human being also appears to be subjected to the law, to the brute force of factual matters. And is this brutality not the strongest enemy of the inner life enclosed in his I, in the being of his soul? All technology is a creation of mind; mind can only ground its own mastery in this way, because it conquers all powers that find themselves enclosed upon it, despotically holding them down. To become master, it must not only restrict the free realm of the soul, it must also deny and destroy it. No compromise is possible in this conflict. Mind, whose goal and power emerges in technology, is the irreconcilable opponent of the soul. And, as it progressively alienates the human being from his own centre of life, the same thing occurs concerning the human relationship to the whole of nature – insofar as this is not taken in one of the senses already distorted by technology, insofar as it is not thought of as a mere mechanism obeying general laws, but rather felt in its organic peculiarity and in its organic fullness of life. The more the power of technology grew within the circles of modern culture, the more passionately and more inexorably relentless did philosophy levy this complaint and accusation against it. As Ludwig Klages, the most eloquent and radical proponent of this fundamental idea, writes: 'Whereas all living creatures except for human beings beat with the rhythm of cosmic life, the human being has severed the law of spirit from this. What appears to him, the bearer of I-consciousness, in light of the superiority of anticipatory thinking over the world, appears to metaphysicians, when they penetrate sufficiently deeply, in light of the enslavement of life under the servitude of concepts. [The human being] has himself fallen out with the planet that bore and nurtured him, even with the cycle of change of all heavenly bodies, because he is possessed by this vampiric and soul-destroying power.'[29]

We miss the actual meaning of these accusations if we believe ourselves able to moderate or overcome them by simply remaining here with the observation of the appearances, with the bare effects. Here it does not suffice to compare the pernicious effects of the rational-technological stance, which are perfectly clear, with other pleasant and beneficial consequences, drawing an acceptable or favourable balance out of this comparison by a 'hedonistic calculus'. For the question is not directed to the consequences but to the ground, not to the events but to the functions. It is from such observation and analysis of function that the critique of a determined cultural content and cultural domain must begin. In the centre of this critique there must always stand the question about the human being himself, about his meaning and 'determination'. In this sense, Schiller, standing at the apex of a particular epoch of aesthetic-humanist culture, poses the question about the significance and value of the 'aesthetic'. And he answers this question by saying that art is not a mere possession and it is no less a mere performance or act of the human being; rather, it must be understood as a necessary path towards becoming human and as a particular phase along this path. It is not the human being who, as mere natural being, as a physical-organic being, becomes the creator of art; rather it is art that proves to be the creator of humanity, that first constitutes and makes possible the specific 'mode' of being human. The ludic drive upon which Schiller grounds the region of beauty does not simply add to the mere natural drives such that it would be a broadening of their range, but rather this drive transforms their specific content, first opening up and conquering the proper sphere of 'humanity'. 'The human only plays where he exists in the genuine meaning of the word "human", and *he is completely human only when he plays.*'[30] This totality of humanity appears to have been realized in no other function in the same sense and to the same measure as in art. We could easily trace how, in German intellectual history, this purely aesthetically composed and grounded 'humanism' gradually grew, and how another cultural power locates itself, independently and equally, next to art. For Herder and Humboldt it is language that shares with art the role of creator and seems to be the basic motive for the real 'anthropogeny'. The domain of technological efficacy seems, however, to be denied any such acknowledgement. For, this efficacy appears to be completely subjected by the mastery of those drives, which Schiller characterizes as the sentient impulse or as the material drive. The urge towards the outside – that typically 'centrifugal' impulse – manifests itself in it. It brings one piece of the world after another under the dominion of the human will; this spread, this expansion of the periphery of being, thereby leads further and further away from the centre of the 'person' and personal existence. Thus it seems that every advance in width must be bought at the cost of a loss of depth. Can it in any way be said of such a function, even if we turn to the most indirect sense of the word that Schiller has stamped on art, that it is not only a creation of the human being, but that it is also his 'second creator'?

Certainly, a general consideration arises against the constitutive interpretation that wants to see technology as an endeavour directed only towards an outside. Here, Goethe's claim that nature has neither core nor shell rightly applies to the totality of mental activities and energies. Here there is no separation, no absolute barrier between the 'outer' and 'inner'. Each new gestalt of the world opened up by these energies is likewise always a new opening out of inner existence; it does not obscure this existence, but makes it visible from a new perspective. We always have before us a manifestation from the inner to the outer and from the outer to the inner – and in this double movement, in this particular oscillation, the contours of the inner and the outer world and their two-sided borders are determined. This is also true for technological efficacy because it is in no way directed towards the seizing of a mere 'outside', but rather it encloses in itself a particular turn inward and backward. Here too it is not about breaking one pole free from another, but rather about both being determined through each other in a new sense. If we move from this determination, then it would appear at first that knowledge of the I is tied in a very particular sense to the form of technological doing. The border that separates purely organic efficacy from this technological doing is likewise a sharp and clear demarcating line within the development of I-consciousness and singular 'self-knowledge'. From the purely physical side, this shows itself in the fact that a determined and clear consciousness of his own body, both a consciousness of his bodily gestalt and his physical functions, first grows in the human being after he turns both of these towards the outside and, so to speak, regains both from the reflection of the outer world. In his *Philosophie der Technik*, Ernst Kapp sought to think through the idea that the human being is granted knowledge of his organs only by a detour through organ-projection. By organ-projection he understands the fact that an individual limb of the human body does not simply work outward, but it creates an outer existence, so to speak, an image of itself. Every primitive work-tool is just such an image of the body; it is a contrary playing-out and reflection of the form and activity of the living body in a determined material structure of the outer world. Likewise, every tool that can be used by the hand appears in this sense as a further laying out and formation, as an exteriorization, of the hand itself. In all its conceivable positions and movements, the hand has provided the organic prototypes after which the human being has unconsciously formed his first necessary pieces of equipment. Hammers and axes, chisels and drills, scissors and tongs are projections of the hand. 'The parts of the hand, its palm, thumb and fingers, the open, hollow, finger-spreading, turning, grasping and clenched hand are, either alone or simultaneously with the stretched or bent forearm, the common mother of the tool named after it.' From this Kapp draws the conclusion that the human being was only able to gain an insight into the composition of his body, into his physiological structure, through the artificial counter-image, through the world of artefacts he himself created. Only insofar as he learned to produce certain physical-technological apparatuses did he truly come to know the structure of his organs in and through them. The eye, for example, was the model for all optical apparatuses. The properties and function of the eye, however, have only been understood through these apparatuses: 'Only as the sight organ had projected itself into a number of mechanical tasks, thus preparing their relation back to its anatomical structure, could this physiological puzzle be solved. From the instrument unconsciously formed according to the organic tool of seeing, the human being has, in a conscious manner, transferred the name to the actual focus of the reflection of light in the eye – the *crystal lens*.'[31]

We cannot closely follow the metaphysical content of this thesis or the metaphysical justification that Kapp has given for it. Insofar as this justification is based upon essentially speculative assumptions, including Schopenhauer's theory of the will and upon Eduard von Hartmann's *Philosophie des Unbewußten*, it is justly disputed and sharply criticized.[32] But this criticism does not destroy the basic perspective and insight Kapp expresses when he says that technological efficacy, when outwardly directed, likewise always exhibits a self-revelation and, through this, a means of self-knowledge.[33] Admittedly, if we assume this interpretation, a radical consequence cannot be avoided – namely, with this first enjoyment of the fruit from the tree of knowledge the human being has cast himself out forever from the paradise of pure organic existence and life. We may with Kapp still attempt to understand and interpret the first human tools as mere continuations of this existence; we may rediscover in the shape of the hammer, axe, chisel, drill and tongs nothing other than the being and structure of the hand itself. If we go one step further, however, and enter into the sphere of advanced technology, this analogy immediately breaks down. This sphere is governed by a law that Karl Marx called the law of the 'emancipation of the organic barrier'. What separates the instruments of fully developed technology from primitive tools is that they have, so to speak, detached and dissociated themselves from the model that nature is able to immediately offer them. What these instruments have to say and what they have to accomplish completely comes to light only because of this 'dissociating'. As to the basic principle that rules over the entire development of mechanical engineering, it has been pointed out that the general situation of machines is such that they no longer seek to imitate the work of the hand or nature, but instead seek to carry out tasks with their own authentic means, which are often completely different from natural means.[34] Technology first attained its own ability to speak for itself by means of this principle and its ever-sharper implementation. It now erects a new order that is not grounded on the contact with nature, but rather, not infrequently, in conscious opposition to it. The discovery of new tools represents a transformation, a revolution of the previous types of efficacy and the mode of work itself. Thus, as other thinkers have emphasized, with the advent of the sewing machine comes a new way of sewing, with the steel mill a new way of smithing – witness the problem of flight, which could only finally be solved once technological thinking freed itself from the model of bird flight and abandoned the principle of the moving wing.[35] Once again, a penetrating and surprising analogy appears here between the technological and linguistic function, between the 'mental aspect of the tool' and the 'the tool of the mind'. For language in its beginning still seeks to hold fast to the 'proximity with nature'. It devotes itself to the direct sense impression of the thing, and then strives to hold on to its sound and, as much as possible, to its sound image, and, in a sense, to exhaust itself in it. But the further it progresses on its way, the more it dissociates itself from this immediate constraint. It abandons the path of onomatopoetic expression; it wrestles itself free from the mere metaphor of sound in order to turn into the pure symbol. And with this it has found and established its own mental gestalt; the power dormant in it has arrived at a true break-through.[36]

Thus, here too the march of technology is mastered by a universal norm that rules the whole of cultural development. The transition to this norm, however, cannot, of course, take place here, as in the other spheres, without struggle and the sharpest opposition. The human being faces the risk of absolving himself from the guardianship of nature, standing purely on his own and on his own wanting and thinking. He has herewith renounced all the benefit that is contained in his immediate proximity to nature. And once the bond that binds him to nature is cut, it can never be tied again in the old way. The moment the human being devotes himself to the hard law of technological work, the abundance of immediate and unbiased happiness that organic existence and activity had given him fades away forever. From the first and most primitive levels it appears as if a close connection still existed between the two forms of efficacy, as if there occurred between them a constant, almost unremarkable transition. Karl Bücher, in his writing on *Arbeit und Rhythmus*, explains how the simplest works accomplished by humanity are still closely connected and related to certain prototypes of the rhythmic movement of one's own body.[37] They appear as the simple continuation of these movements; they are not so much directed by a determined idea of an external goal as they are inwardly motivated and determined. What is represented in these works and what directs and regulates them is not a goal-conscious will, but a pure impulse to expression and naive joy of expression. Even today this connection can be directly detected in the widespread customs of native peoples. It is reported that in many indigenous tribes dance and work are denoted by the same word. Both are for them phenomena so immediately related and so insolubly bound together that they cannot linguistically and intellectually be distinguished from one another. The success of agricultural labour depends not only upon certain external technological performances but also upon the correct execution of their cultural chants and dances; it is one and the same rhythmic movement that both forms of activity enclose, bringing them together into the unity of a singular, unbroken feeling of life.[38] This unity appears immediately endangered and threatened as soon as activity takes the form of indirectness, as soon as the tool comes between the human being and his work. For the tool obeys its own law, a law which belongs to the world of things, and which, accordingly, breaks into the free rhythm of natural movements with a foreign dimension and foreign norm. The organic bodily activity asserts itself over and against this disturbance and inhibition insofar as it manages to include the tool itself in the cycle of natural existence. This inclusion still appears to succeed without difficulty at the relatively early stages of technological work activity. Organic unity and organic connection reinstate and reproduce themselves insofar as the human being continues to 'grow together' with the tool he employs, so long as he does not look upon the tool as merely stuff, a mere thing composed of matter, but instead relocates the tool into the centre of its function and, by virtue of this shifting of focus, feels a kind of solidarity with it. It is this feeling of solidarity that animates the genuine craftsman. In the particular individual work that is created by his hands he has no mere thing before him; in it he sees both himself and his own personal activity. The further the technology progresses and the more the law of 'emancipation from the organic barrier' affects it, the more this original unity slackens until it finally breaks up completely. The connection of work and working ceases in any way to be a connection one can experience, because the end of working, its proper telos, is now entrusted to the machine, while the human being essentially becomes, in the whole of the process of work, something dependent – a section or part that is increasingly converted into a mere fragment. Simmel sees the essential reason for what he calls the 'tragedy of modern culture'[39] in the fact that all creative cultures increasingly set out certain orders of things for themselves that confront the world of the I in their objective existence and in their being-such-and-such. The 'I', the free subjectivity, has created these orders of things, but it no longer knows how to grasp these things and how to penetrate into them. The movement of the 'I' breaks upon its own creations; the greater the scope and the stronger the power of this creation becomes, the more its original tide of life subsides. This tragic element of all cultural development is perhaps no more evident than in the development of modern technology. But those who turn away from it on the basis of these findings forget that in their damning judgement of technology they must logically include the totality of culture. Technology has not created this state of affairs. It merely places an especially remarkable example urgently before us. It is – if one speaks here of suffering and sickness – not the ground of suffering, but merely a manifestation, a symptom of it. What is crucial here is not an individual sphere of culture but its function, not a special way that it follows, but the general direction it takes. Thus, technology may at least demand that the charges raised against it not be brought before the wrong court. The standard by which it alone can be measured can, in the end, be none other than the standard of mind, not that of mere organic life. The law that one applies to it must be taken from the whole of the mental world of forms, not merely from the vital sphere. Thus grasped, however, the question as to the value and demerit of technology immediately receives another sense. It cannot be resolved simply because one considers and sets off against each other the 'utility' and 'disadvantages' of technology. We cannot judge it by comparing the good that it gives to humankind with the idyll of some pre-technological 'state of nature'. Here, it is about neither pleasure nor displeasure, neither happiness nor sorrow. It is about freedom and bondage. If the growth of technological ability and wares necessarily and essentially secures in itself a stronger measure of servitude such that it increasingly enslaves and constrains humanity rather than being a vehicle for its self-liberation, then technology is condemned. If the reverse shows itself – that is, if it is the idea of freedom itself that shows the way for technology and finally breaks through in it – then the significance of this goal cannot be curtailed by looking at the suffering and troubles technology causes along the way. For the path of mind stands here as everywhere under the law of renunciation, under the command of a heroic will that knows it can only reach its goal through such renunciation, establishing itself through it and renouncing all naive and impulsive longings for happiness.

IV

The conflict generated between the human longing for happiness and the demands imposed on it by the technological mind and technological will is, however, in no way the sole and strongest opposition that emerges here. The conflict becomes deeper and more menacing when it emerges in the sphere of cultural forms. The true battlefront first appears where the mediating mind no longer merely struggles with the immediacy of life, but when the mental tasks become increasingly differentiated and simultaneously alienate themselves further from one another. For now, it is not only the organic unity of existence, but also the unity of the 'idea', the unity of direction and purpose, which are threatened by this alienation. Moreover, as technology unfolds, neither does it simply place itself next to other fundamental mental orientations nor does it order itself harmoniously and peacefully with them. Insofar as it differentiates itself from them, it both separates itself from them and positions itself against them. It insists not only on its own norm, but also threatens to posit this norm as an absolute and to force it upon the other spheres. Here, a new conflict erupts within the sphere of mental activity, indeed, on its very lap. What is now demanded is no simple confrontation with 'nature', but the erection of a barrier within mental life itself – a universal norm that both satisfies and restrains individual norms.

The determination of this barrier is most easily fixed in technology's relation to the theoretical knowledge of nature. Here, harmony seems to be given and guaranteed from the beginning. There is no struggle for superiority and subordination, but a reciprocal giving and taking. Each of the two basic orientations stands on its own. However, even this independence unfolds freely and spontaneously in an unforeseen manner towards a pure subservience to and with the other. The truth of Goethe's words – that doing and thinking, thinking and doing, constitute the sum of all wisdom – appears nowhere more clearly than here. For it is in no way the 'abstract', pure theoretical knowledge of the laws of nature that leads the way, proving first the technological aspect of the problem and its concrete technological activity. From the very beginning, both processes grasp one another and, as it were, keep the balance. Historically, this connection can be made clear when we look back at the 'discovery of nature' that has taken place in European consciousness since the days of the Renaissance. This discovery is in no way the work of only the great researchers of nature – it returns essentially to an impulse originating out of the questions of the great inventors. In a mind like that of Leonardo da Vinci the intertwining of these two basic orientations appears with a classic simplicity and depth. What separates Leonardo from mere bookish learning, from the spirit of *'letterati'*, as he himself called it, is the fact that 'theory' and 'praxis', 'praxis' and 'poiesis', penetrate one another in his person in a completely different measure as never before. First an artist, he became a technician and then a scientific researcher. Likewise, for Leonardo all research transforms directly into technological problems and artistic tasks.[40] This is hardly a question of a mere one-time connection but rather of a factual and basic connection that, from here onwards, points the way for the entire science of the Renaissance. The actual founder of theoretical dynamics, Galileo, also began from technological problems. In his book on Galileo, Olschki rightly places the strongest emphasis on this element. He notes that 'very few of the biographies have directed attention to this side of Galileo's work and scientific development. To be more precise, however, this more original and persistent of his varied dispositions constituted the main focus of his seemingly disparate life works ... One must keep in mind the fact that each of Galileo's discoveries in physics and astronomy are closely linked to some instrument of his own invention or to some special set-up. His technological genius is the authentic prerequisite for the scientific efforts through which his theoretical originality first received its direction and expression.'[41] The genuine explanation of these facts is that theoretical activity and technological activity do not only touch one another externally, insofar as they both operate on the same 'material' of nature, but, more importantly, they relate to one another in the principle and core of their productivity. The image of nature that thought produces is not captured by a mere idle beholding of the image; it requires the use of an active force. The more one steeps oneself in critical epistemological reflection about the origins and conditions of this image, the more it becomes clear that this image is no simple copy – that its outline is not simply drawn from nature – but that it must be formed from an independent energy of thought. Here we have arrived at the point where reason, according to Kant, appears as the 'author of nature'. This authorship, however, assumes another direction and attests to a new path as soon as we consider the workings of technological creation. Technological work and theoretical truth share a basic determination in that both are ruled by the demand for a 'correspondence' between thought and reality, an 'adaequatio rei et intellectus'.[42] That this 'correspondence' is not immediately given, but rather is to be searched for and continuously produced, appears even more clearly in technological creation than in theoretical knowledge. Technology submits to nature in that it obeys its laws and considers them as the inviolable requirements of its own workings. Notwithstanding this obedience towards the laws of nature, however, nature is never for technology something finished, wherein laws are merely posited. Nature is something that is to be perpetually posited anew, something that is to be formed repeatedly. Mind always measures anew objects in relation to itself, and itself in relation to objects, in order to find and guarantee in this twofold act the genuine *adaequatio*, the actual 'appropriateness', of both. The more this movement takes hold, the more its force grows, the more the mind feels and knows its reality to have 'grown'. This inner growth does not simply take place under a continuous leadership, under the rule and guardianship of the actual; rather, it demands that we constantly return from the 'actual' to a realm of the 'possible', and see the actual itself according to this image of the possible. Acquiring this point of view and orientation signifies, from a purely theoretical perspective, perhaps the greatest and most memorable achievement of technology. Standing in the middle of the sphere of necessity and remaining within the idea of necessity, it discovers a sphere of free possibilities. There is no uncertainty, no mere subjective insecurity attached to these possibilities; they confront thought as something thoroughly objective. Technology does not initially ask what is but what can be. This 'ability', however, designates no mere assumption or supposition, but an assertive claim and certainty – a certainty whose final authentication, of course, is to be sought not in mere judgement, but in the output and production of certain artefacts. In this sense, every truly original technological achievement has the character of both a discovering and an uncovering. A certain state of affairs is in a sense extracted from the region of the possible and transplanted into the actual. Here, the technician bears a likeness to the activity of the divine 'demiurge' in Leibniz's metaphysics who does not create the essence or possibility of objects, but selects only one, and the most perfect, among those possibilities that exist in themselves and are presently at hand. Thus technology repeatedly teaches us that the sphere of the 'objective', which is determined by fixed and general laws, never coincides with the sphere of that which is presently at hand – that is, with that which becomes actual through the senses.[43] Pure theoretical natural science can, of course, never know the actual without constantly reaching out into the realm of the possible, the purely ideal. In the end, however, the only actuality to which its gaze appears to be directed seems to have exhausted itself in the clear and distinct description of the actual processes of nature. Technological work, however, never binds itself to this pure facticity, to the given face of objects; rather it obeys the law of a pure anticipation, a prospective view that foresees the future, leading up to a new future.

With the insight into this state of affairs, however, the authentic centre of the world of technological 'form' now seems to shift increasingly, and to cross over from the pure theoretical sphere into the sphere of art and artistic creation. Here, we need not prove how tightly both areas are interwoven with one another. A glance at general intellectual history suffices to teach us how fluid the transitions are in the concrete becoming, in the genesis, of the technological world of form and in artistic form. Again, the Renaissance, with its construction of the *'uomo universale'*[44] in such spirits as Leon Battista Alberti and Leonardo da Vinci, provides us with great examples of the constant interweaving of technological and artistic motives. Nothing appears more natural and more enticing than concluding that such a coincidence in fact can come from such a coincidence in person. Indeed, there are those among the modern apologists of technology who believe that they can serve their cause in no better way than by equating it with the cause of art. They are, as it were, the romantics of technology. They attempt to ground and justify technology by dressing it up with all the magic of poetry.[45] All poetic hymns about the achievements of technology cannot, of course, raise us above the task of determining the difference between technological and artistic creation. This difference immediately emerges if we consider the kind of 'objectification' that is actual in the artist and in the technician.

In the present-day literature on the 'philosophy of technology', we repeatedly encounter the questions of whether and to what extent a technological work is capable of producing pure aesthetic effects and to what extent it is subject to pure aesthetic norms. The answers given to these questions are diametrically opposed to one another. The 'beautiful' is quickly claimed and praised as an inalienable good of technological products, and just as quickly rejected as a 'false tendency'. This struggle, often fought with great bitterness, wanes when one considers that in the thesis and antithesis the concept of beauty is, for the most part, taken in an entirely different sense. We grasp the norm of 'beauty' so widely that we speak of it everywhere there emerges a victory of 'form' over 'stuff', 'idea' over 'matter', such that there can be no doubt as to the great extent of technology's direct role. This beauty of form encompasses par excellence the whole expanse of mental activity and formation in general. Understood in this sense, there is, as Plato said in the *Symposium,* not only a beauty of physical formation but also of logic and ethics, a 'beauty of knowledge' and a 'beauty of custom and endeavors'.[46] To reach the special region of artistic work from this all-embracing concept of form an essential limitation and a specific regulation are required. This results from that original relation in which all artistic beauty stands in relation to the grounding and original phenomenon of expression. In an absolutely unique way that is reserved for it alone, the work of art permits 'gestalt' and 'expression' to merge into one another. It is a creation that reaches out into the realm of the objective and that places before us a rigorous objective lawfulness. However, this 'objective' is in no way a mere 'appearance'. It is the expression of something interior and gives a certain transparence to it. The poetic, painted, or plastic form is in its highest perfection, in its pure 'detachment' from the 'I', still flooded by the pure movement of the 'I'. The rhythm of this movement lives on mysteriously in the form and speaks to us immediately in it. The outline of the gestalt turns back here repeatedly to a certain trait of the soul that manifests itself in it; and, in the end, it is to be rendered understandable only from the whole of this soul, from its totality that is enclosed in each true, artistic, individual thing. Such wholeness and such individual particularity continue to be denied to technological work. Admittedly, if one restricts oneself only to the mere experiential content of technological and artistic creations, then there appears to be no strict border between the two. Indeed, when it comes to intensity, fullness and passionate emotion, the one is not inferior to the other. And when the work of a discoverer or inventor first breaks through into reality after years and years of being carried inwardly, it involves no less a psychical or mental tremour than when the poetic or plastic gestalt detaches itself from its originator, confronting him as a figure in its own right. But after this separation has taken place even once, a quite different connection between the creator and his work prevails in the purely technological sphere as compared to the artist and his work. The completed object, in becoming actual, belongs to reality. It is situated in a pure world of things whose laws it obeys and by whose measure it wants to be measured. It must henceforth speak for itself, and it speaks only of itself and not of the creator to whom it originally belonged. This type of detachment is not demanded of the artist and it is not possible for him. Even when he becomes completely absorbed in his work, he does not become lost in it. The work always remains – insofar as it stands purely on its own – simultaneously the testimony of an individual form of life, an individual Dasein and a particular kind of being. Technological creation can neither reach nor aspire to reach this sort of 'harmony' between the beauty of the work and the beauty of expression. When, with the erection of the Eiffel tower, the artists of Paris united and rallied in the name of artistic taste to object to this 'useless and monstrous' construction, Eiffel answered them that he was firmly convinced that his work had its own beauty: 'Are the right conditions of stability not always in agreement with those of harmony? The foundation of the art of building is that the main lines of the building must completely correspond to certain rules. What is, however, the basic condition of my tower? Its resistance against the wind! And here I claim that the curve[s] of the four pillars of the tower that climb higher and higher into the air in accordance with the fixed measurements of the weight of the base make for a powerful impression of force and beauty.'[47] This beauty, which originates from the perfect solution to a given problem, is, however, not of the same type and origin as the beauty that confronts us in the work of poets, sculptors and musicians. This latter beauty is not based on 'being bound' by the forces of nature, but also represents a new and unique synthesis of the 'I' with the world. If we can denote the world of expression and the world of pure signification as the two extremes between which all cultural development moves, then the ideal balance between them is, as it were, achieved in art. Technology combined with theoretical knowledge, to which it is closely related, renounces increasingly all that is measured by expression in order to lift itself up into the strictly 'objective' sphere of pure meaning.[48] At the same time, it is indisputable that the gain achieved here contains a sacrifice. But even this sacrifice and this renunciation, this possibility to cross over and rise up into a pure world of things, shows itself to be a specific human power – an independent and indispensable descriptor of 'humanity'.

However, a deeper and more serious conflict erupts before us if, rather than measuring technological works and activity by aesthetic norms, we ask after its ethical right to exist and its ethical meaning. The moment this question is vigorously put forth and understood in its entire severity, the decision seems already to be made. The sceptical and negative critique of culture, which Rousseau introduced in the eighteenth century, seems to be able to give no weighty evidence, no stronger example than the development of modern technology. Does this development not, under the promise and alluring image of freedom of the travelling juggler, involve human beings even more inexorably in bondage and enslaved? In that it removes him from the bond to nature, has it not increased his social bonds to the point of being unbearable? The thinkers who have struggled most profoundly with the basic problem of technology are precisely those who have repeatedly enjoyed this ethically damning judgement over it. Whoever does not from the beginning subscribe to the demands of simple utility, and instead treasures the meaning of ethical and spiritual standards, cannot carelessly pass over the grave inner damages of a lauded 'technological culture'. Few modern thinkers have as keenly observed and forcefully uncovered this damage as Walther Rathenau.[49] He has done so with growing zeal and passion in his writing. On one hand, there is completely soulless and mechanized work, the hardest chore. On the other hand, there is unrestricted will to power and will to rule, unrestrained ambition and meaningless consumerism. Such is for Rathenau the picture of the times captured in the mirror of technology: 'If one considers ... world production, the insanity of the economy appears to us terribly frightening. Superfluous, trivial, harmful, contemptuous things are heaped in our stores, useless fashion statements that should, in a few days, emit a false radiance, ways of getting intoxicated, stimulus, a numbing ... Every new financial quarter, all these worthless things fill stores and warehouses. Their manufacture, transportation and consumption require the work of millions of hands; they demand raw materials, machines, plants, occupying approximately one-third of the world's industry and workers.'[50] Modern technology, and the modern economy, which has apparently created and sustained itself by its own means, is the true jug of the Danaides. This image, already used by Plato in *Gorgias* to describe the vanity and absurdity of an ethics measured according to purely hedonistic criteria, spontaneously forces itself upon us when we read Rathenau's description. Every satisfied need serves only to bring forth new needs in increasing measure – and, once you have entered it, there is no escape from this cycle. Seizing the human being even more relentlessly than the workings of his own drives is the working of the drives of his situation that is the result and product of technological culture; he is thrown by technological culture into a never-ending vertigo that moves from desire to consumption, from consumption to desire.

As long as we remain in the sphere of its external appearance, its consequences and effects, the hard verdict cast here upon technology is without appeal. Only one question can still be asked: whether these effects can necessarily be attributed to its essence, that is, whether they are enclosed in the principle of formation of technology, and whether they are demanded by it. When the problem is taken in this sense, a thoroughly different aspect of the observation and judgement emerges. Rathenau leaves no doubt that all the gaps and damage of modern technological culture he inexorably uncovers not only come from itself, but rather have to be understood in terms of their connection with a certain form and order of commerce. Every attempt at improvement must begin here. This connection does not originate in the culture of technology. It is more the case that it is made necessary and thrust upon one by a particular situation, by a concrete historical position.[51] Once this interconnection is established, however, it cannot be undone by means of technology alone. It is not enough here to appeal to the forces of nature or mere understanding to technological and scientific intellects. Here it suffices to indicate the point at which only the deployment of a new willpower can create change. In this construction of the realms of will and the basic convictions upon which all moral community rests, technology can only ever be a servant, never a leader. It cannot by itself determine the goal, although it can and should collaborate in carrying it out. It best understands its own meaning and its own narrative when it is content in the fact that it can never be an end itself. Rather, it has to fit itself into another 'realm of purpose', into a genuine and final teleology that Kant described as ethico-teleological. In this sense, the 'dematerialization', the 'ethicization' of technology forms one of the central problems of our present culture.[52] Just as technology could not immediately create ethical values out of itself and its own circle, there cannot exist alienation and opposition between technology's values and its specific direction and basic convictions. This is the case because technology is governed by 'practical thinking', by the idea of a solidarity of work in which all ultimately work for one and one works for all. It creates – even before the truly free community of wills – a sort of community of fate between all those who are active in its work. Thus, we can correctly define the implicit meaning of the technological world and technological culture as the idea of 'freedom through bondage'.[53] If this idea is truly to have an effect, it is, of course, necessary that it transform more and more its implicit meaning into an explicit one. That which appeared in technological creation is recognized and understood in its basic direction, that it is raised into mental and moral consciousness. Only if this happens does technology prove not only to be the vanquisher of the forces of nature but also the vanquisher of the chaotic forces of the human being. All the defects and failings one is in the habit of advancing today are, in the end, based upon the fact that until now it has not fulfilled its highest mission. In fact, it has hardly yet recognized it. All 'organization' of nature, however, remains questionable and sterile, provided that it does not lead to the goal of the formation of the will to work and the real and fundamental work attitude. Still, our culture and our present society are far from this goal. Only when this is understood as such and methodically and energetically grasped, will the real relationship between 'technology' and 'form', its deepest form-forming strength, be able to prove itself.

Translated by Wilson McClelland Dunlavey and John Michael Krois

Notes

1. First published in *Kunst und Technik*, ed. Leo Kestenberg (Berlin: Wegweiser, 1930), 15–61. Our translation is based both on the essay's first publication in *Kunst und Technik* and the later publication in Ernst Cassirer's *Gesammelte Werke*, ed. Birgit Recki (Hamburg: Meiner, 1998–2009) (hereafter cited as ECW), vol. 17 *Aufsätze und kleine Schriften (1927–1931)*, ed. Tobias Berben (2004), 139–83. We have translated the footnotes as they appear in the first edition, as well as some helpful footnotes from the second edition (shown in square brackets). We have followed the typography of the latter edition. Further comments added by us are marked as translators' note.
2. The German word is 'Geist', which is sometimes rendered as 'Spirit' in English, particularly in reference to Georg Wilhelm Friedrich Hegel. 'Mind' is our preferred term, though the 'mind' of René Descartes should not be presupposed (translators' note).
3. [Georg Wilhelm Friedrich Hegel, *Grundlinien der Philosophie des Rechts*, in *Werke*, ed. Eduard Gans, vol. VIII (Berlin, 1833), 19f.]
4. Friedrich Dessauer, *Philosophie der Technik: Das Problem der Realisierung* (Bonn, 1927), 146.
5. Eberhard Zschimmer, *Philosophie der Technik: Vom Sinn der Technik und Kritik des Unsinns über die Technik* (Jena, 1914), 28.
6. Art, skill (translators' note).
7. Plato, *Cratylus* 389 A (for detail, see my presentation of the history of Greek philosophy: 'Die Philosophie der Griechen von den Anfängen bis Platon', in Max Dessoirs (ed.), *Lehrbuch der Philosophie*, vol. I, 7–138, 92f [ECW 16, 313–467]).
8. Dessauer, *Philosophie der Technik*, 150; Max Eyth, 'Poesie und Technik', in *Lebendige Kräfte: Sieben Vorträge aus dem Gebiete der Technik*, 4th edn (Berlin, 1924), 1f.
9. In the scope of this work I can only state this thesis: for the development and the systematic justification of this claim I refer the reader to my *Philosophie der symbolischen Formen* (vol. 1: *Die Sprache*, Berlin 1923 [ECW 11], vol. 2: *Das mythische Denken*, Berlin 1925 [ECW 12] and vol. 3: *Phänomenologie der Erkenntnis*, Berlin 1929 [ECW 13]).
10. [Jean-Jacques Rousseau, *Discours qui a remporté le prix a l'Academie de Dijon, en l'année 1750. Sur cette question proposée par la même Academie: Si le rétablissement des sciences et des arts a contribué à épurer les mœurs*, in: *Collection complette des œuvres*, vol. XIII (Zweibrücken, 1782), 27–62.]
11. ['Do not laugh, do not lament, do not even hate; rather, understand.' Baruch de Spinoza, *Tractatus politicus*, in *Opera postuma* (Hamburg, 1677), 265–354, 268. See, for example, the disparate judgements over the meaning and worth of technology, which Zschimmer has summarized in his *Philosophie der Technik*, 45ff and 136ff.]

12. Eyth, 'Poesie und Technik', 12ff; see also the lecture 'Zur Philosophie des Erfindens', in *Lebendige Kräfte*, 229–64.
13. For details about this 'analogical character of logos' in the theory of language of the sophists, see the explanation of Ernst Hoffmann, *Die Sprache und die archaische Logik* (Tübingen, 1925), 28ff.
14. [Verified by James Boswell (conversation on 7 April 1778), in *Life of Samuel Johnson*, vol. II (London, 1900), 422–30, 425.]
15. Wilhelm von Humboldt, 'Ueber die Verschiedenheit des menschlichen Sprachbaues und ihren Einfluss auf die geistige Entwicklung des Menschengeschlechts', in *Werke*, ed. Albert Leitzmann, vol. VII, part I, (Berlin, 1907), 119; for details, see my *Philosophie der symbolischen Formen*, vol. I: *Die Sprache* (Berlin, 1923). [ECW 11]
16. [Wilhelm von Humboldt, *Werke*, vol. IV, 1–34, 27.]
17. Theodor-Wilhelm Danzel, *Kultur und Religion des primitiven Menschen: Einführung in Hauptprobleme der allgemeinen Völkerkunde und Völkerpsychologie* (Stuttgart, 1924), 2ff, 45ff and 54ff.
18. For a more detailed argument, see the introduction to my *Philosophie der symbolischen Formen*, vol. I, 6ff [ECW 11, 4ff].
19. See James George Frazer, *The Golden Bough: A Study in Magic and Religion*, part I in vol. II: *The Magic Art and the Evolution of Kings* (London, 1911), chapters 3 and 4.
20. Rich ethnological material for this fundamental view can be found in Lucien Lévy-Bruhl, *Das Denken der Naturvölker* (German translation: Wien, 1921).
21. The dominion of man on earth (translators' note).
22. [Francis Bacon, *Novum Organum*, in Robert Leslie Ellis, James Spedding and Douglas Denon Heath (eds), *Works*, vol. I (London, 1858), 70–223, 157.]
23. The middle stage in a process (translators' note).
24. For details, see *Philosophie der symbolischen Formen*, vol. III, 317ff [ECW 13, 315ff].
25. 'Wherever sympathetic magic occurs in its pure unadulterated form, it assumes that in nature one event follows another necessarily and invariably without the intervention of any spiritual or personal agency. Thus its fundamental conception is identical with that of modern science. The magician does not doubt that the same causes will always produce the same effects. Thus the analogy between the magical and the scientific conceptions of the world is close. In both of them the succession of events is perfectly regular and certain, being determined by immutable laws, the operation of which can be foreseen and calculated precisely; the elements of caprice, of chance and of accident are banished from the course of nature.' (Frazer, *The Magic Art*, vol. I, 220f.)
26. Ludwig Noiré, in his book *Das Werkzeug und seine Bedeutung für die Entwickelungsgeschichte der Menschheit* (Mainz, 1880), has emphasized that the particular signification of the work-tool, in its purely intellectual sense, lies in the fact that it represents a basic means in the process of 'objectivation' out of which the worlds of 'language' and 'reason' emerge. 'The great importance of the work-tool', he emphasizes, 'lies mainly in two things: in the solution or singling out of causal relations whereby the latter receives in human consciousness an ever growing clarity and, secondly, in the objectivation or the projection of his organs that had up to now taken place only

in the darkness of the consciousness of an instinctual function.' (34). This thesis remains valid even if one does not agree with the justification given by Noiré – a justification that is founded mainly on linguistic-historical facts and on a certain theory about the origin of the language.

27. For details, see my work *Sprache und Mythos: Ein Beitrag zum Problem der Götternamen* (Studien der Bibliothek Warburg, vol. 6) (Leipzig, 1925), 48ff, 68f [ECW 16, 227-311, 278ff, 298f].

28. [Johann Wolfgang von Goethe, 'Paralipomena', in *Werke* (Weimar, 1887–1919), part. I, vol. XLVII, 275–389, 323.]

29. Ludwig Klages, *Vom kosmogonischen Eros* (München, 1922), 45 (first quote); see also *Mensch und Erde: Fünf Abhandlungen* (München, 1920), 40ff (second quote: 40f).

30. [Friedrich Schiller, *Über die ästhetische Erziehung des Menschen in einer Reihe von Briefen* (1793–4), in *Philosophische Schriften*, vol. II (Stuttgart and Berlin, 1905), 3–120, 59.]

31. Ernst Kapp, *Grundlinien einer Philosophie der Technik: Zur Entstehungsgeschichte der Cultur aus neuen Gesichtspunkten* (Braunschweig, 1877), 41ff, 76ff and 122ff (quotes: 41 and 79).

32. See, for example, Max Eyth, 'Zur Philosophie des Erfindens', 234ff; Zschimmer, *Philosophie der Technik*, 106ff.

33. Kapp, *Grundlinien*, 26.

34. See Franz Reuleaux, *Theoretische Kinematik: Grundzüge einer Theorie des Maschinenwesens* (*Lehrbuch der Kinematik*, vol. I) (Braunschweig, 1875).

35. For details, see Dessauer, *Philosophie der Technik*, 40 ff; Zschimmer, *Philosophie der Technik*, 102ff.

36. For details, see my *Philosophie der symbolischen Formen*, vol. I, 132ff [ECW 10, 133ff].

37. Karl Bücher, *Arbeit und Rhythmus*, 2nd edn (Leipzig, 1899), esp. 24ff.

38. For details, see Preuß, *Religion und Mythologie der Uitoto: Textaufnahmen und Beobachtungen bei einem Indianerstamm in Kolumbien, Südamerika* (Göttingen and Leipzig, 1923), vol. I: *Einführung und Texte*, 123ff, as well as Preuß's essay 'Der Ursprung der Religon und Kunst', *Globus* 86 (1904), 321–7, 355–63, 375–9, 388–392; and *Globus* 87 (1905), 333–7, 347–50, 380–4, 394–400 and 413–19.

39. [See Georg Simmel, 'Der Begriff und die Tragödie der Kultur', in *Philosophische Kultur: Gesammelte Essais* (Potsdam, 1923), 236–67.]

40. For details, see my *Individuum und Kosmos in der Philosophie der Renaissance* (Studien der Bibliothek Warburg, vol. 10) (Leipzig, 1927) [ECW 14, 1–220].

41. Leonardo Olschki, *Geschichte der neusprachlichen wissenschaftlichen Literatur*, vol. III: *Galilei und seine Zeit* (Halle, 1927), 139f.

42. Adequateness of thing and intellect (translators' note).

43. In his *Philosophie der Technik* (47f) Dessauer keenly and poignantly remarks: 'The reunion of an inventor with the object that he has "brought forth" for the first time, is an encounter of unprecedented power, a strong revelation. The inventor looks at that which was achieved by his work, though not by it alone, not with a "I have made you", but with a "I have found you". You were already somewhere; I had to search long for you ... That you did not exist before owes to the fact that it was only now that I found that this is how you are. You could not appear or fulfil your purpose *until you were so in*

my look, as you are in yourself, because this is the only way you can be! Certainly, now you exist in the visible world. But I have found you in another world, and for a long time you refused to cross over into the visible realm, just until I rightly saw your true Gestalt in that other realm.'

44. Universal man (translators' note).

45. One thinks here in particular of the essay by Eyth, 'Poesie und Technik', 9ff.

46. [Plato, *Symposium*, 210 C (*'ἐπιστημῶν κάλλος'*) and 211 C (*'τά κάλά ἐπιτηδένματα'*).]

47. Quoted after Julius Goldstein, *Die Technik* (*Die Gesellschaft: Sammlung sozial-psychologischer Monographien*, ed. Martin Buber, vol. 40) (Frankfurt am Main, 1912), 51.

48. Concerning theoretical knowledge, this process is explained and developed further in my *Philosophie der symbolischen Formen*, vol. III, part III, 472–559 [ECW 13, 468–556].

49. See esp. Walther Rathenau, *Zur Kritik der Zeit* (Berlin, 1912), and, by the same author, *Zur Mechanik des Geistes oder vom Reich der Seele* (Berlin 1913) and *Von kommenden Dingen* (Berlin, 1917).

50. [Rathenau, *Von kommenden Dingen*, 91f.]

51. Concerning the necessary disjunction between technology and economy, see, in addition to the writings of Rathenau, the remarks by Zschimmer, *Philosophie der Technik*, 154ff, and Dessauer, *Philosophie der Technik*, 113ff.

52. The problem of this 'ethicization' is rightly emphasized by Viktor Engelhardt, *Weltanschauung und Technik* (Leipzig, 1922), 63ff, and by Richard Nikolaus Coudenhove-Kalergi, *Apologie der Technik* (Leipzig, 1922), 10 ff.

53. Dessauer, *Philosophie der Technik*, 86; see esp. 131ff.

2
The Age of Complete Mechanization

John Michael Krois

The ability to make and use tools is one of the defining characteristics of human beings, but, until the twentith century, philosophers gave this fact little heed. Human beings were defined as 'rational animals', so that the life of contemplation, Aristotle's *bios theoretikos*,[1] rated higher than a life of doing or making. Aristotle wrote about tragedy, but not about sculpture, for, unlike dramatists, sculptors were considered to be artisans, engaged in dirty manual work. As Erwin Panofsky commented in his famous study of ancient theories of art, being a sculptor must have been considered to be something especially crude – 'etwas besonders Banausisches in ancient Greece'.[2] Artists were not considered to be creators. In Plato's dialogues, artists are compared to people going around holding up a mirror, and – without any knowledge – reproducing everything: the Sun, and everything in the heavens and on earth.[3]

The ancient view that art (and technical activity) were imitative held sway for centuries, but it no longer made sense at the beginning of the twentieth century. By then, technology had not only assumed new prominence, it had undergone a transformation. Technology could no longer be considered to just extend what was already there, tools were not just mere extensions of the body, and the objects created by technical means no longer had models in nature. This is what the philosopher Ernst Cassirer claimed in his essay 'Form and Technology'[4] in 1930. Flight became possible for humans, he wrote, only by creating unmovable wings, which do not occur in nature.[5]

In the 1920s the question of technology was a matter of general concern, not just among intellectuals. By 1930 every aspect of life, from the workplace to the kitchen and living room, was suddenly affected by new technologies. When the architecture historian Siegfried Gideon published his now classic book *Mechanization Takes Command*[6] in 1948 he called the decade between 1920 and 1930 the age of 'complete mechanization', a time when every sphere of daily life came under the effect of a host of technical innovations. Animals disappeared as a means of transport, domestic electricity became widespread, as did the telephone and phonograph. Few technical innovations, however, were able to enter the private sphere as dramatically as radio. Suddenly, the distant outside world could be experienced in people's private homes. The first regular radio broadcasts for entertainment began in Germany on 29 October 1923 in Berlin with a one-hour programme called the 'Funkstunde', which broadcast music during its first evening. It would not be an exaggeration to say that this event is what eventually led to the publication of Ernst Cassirer's essay on technology in 1930.

It was no coincidence that Ernst Cassirer's essay on technology appeared when and where it did, in a volume published in Berlin devoted to the topic of 'art and technics' (*Kunst und Technik*). The book focused upon the ways in which technical innovations were affecting traditional performance arts, particularly music. Its editor, Leo Kestenberg (1882–1962),[7] was deeply concerned about these matters but he was also in a position to do something about them. Kestenberg was originally a classically trained pianist with a profound interest in music education, but the political changes in Germany following the 1918 revolution enabled him to become a professor of music at the famous Berlin Conservatory, the Akademische Hochschule für Musik, and even a member of the Prussian Ministerium für Wissenschaft, Kunst und Volksbildung. There he headed the section dealing with music (as 'Referent für musikalische Angelegenheiten'). Kestenberg is best remembered for his influential efforts to reform music education in Germany, now known as the 'Kestenberg-Reform'. Kestenberg's aim was to have music education begin early, in kindergarten, and to continue throughout the entire length of the school years. He changed the programme from the teaching of 'singing' to 'music' generally. His goal was to make music enjoyable to all by creating a universal musical literacy. Ideally, this would permit everyone to enjoy even the most advanced and still unpopular kinds of music. These goals were guided by Kestenberg's socialistic humanism, which in practical terms meant that he envisioned everyone having the opportunity to hear the finest artists perform the best music. Kestenberg furthered 'new music' by supporting the hiring of Paul Hindemith and Arnold Schoenberg as professors of composition in Berlin, and he encouraged the experimental Berlin Kroll-Oper, which offered now-legendary modern stagings of modern opera.[8] Kestenberg recognized early that radio was going to change music, and that art was entering a new era with the advent of the new technical forms emerging in the 1920s. Given all this, it is not surprising that Kestenberg solicited papers from experts on different aspects of the effects of technologies on the performing arts and society in general.

To get an idea of the situation: the studios used for the first regular broadcasting in Germany were housed in rooms in a building in downtown Berlin, the Vox-Haus, at Potsdamer Strasse 4. Only five years later, when a new building was constructed especially for radio studios – the 'Haus des Rundfunks' – it had 14,361 square metres (154,580 square feet) of space and took in an entire city block. 'Listening to the radio' was a new activity, which spread from 1,025 paying listeners for the Berlin station in its first year to 220,592 the next, with continuing exponential growth thereafter. From the beginning, music was one of the main contents of radio broadcasting in Germany.

Ernst Cassirer too was in a position to understand what was occurring in the age of 'total mechanization', especially radio. The Cassirer family was part of this technical revolution. Ernst Cassirer's cousin Hugo Cassirer was in charge of engineering at the family's Berlin factory, 'Dr Cassirer & Co. Kabel- und Gummiwerke'. The company produced shielded wire cables for the clear reception of radio broadcasts from antennas, which would permit reception of radio signals undisturbed by interference from electric motors or other such devices.[9] The firm was so successful that by the end of the 1920s the factory moved into a new, larger factory building, built in 1928–29 by the well-known architect Hans Poelzig.[10] Poelzig also designed the aforementioned 'Haus des Rundfunks' from which radio programmes still originate.[11] Ernst Cassirer also gained firsthand experience in broadcasting, probably due to his longtime friend, the philosopher Hans Reichenbach, who had worked for many years as an engineer in the field of radio, first in the military and then in Berlin in the laboratory of a radio firm. Reichenbach published a series on radio broadcasting to which he himself also contributed.[12] He also organized educational radio broadcasts in Berlin with professors from different fields, and it was probably Reichenbach who set Cassirer in touch with the Deutschlandsender.[13]

Beginning in 1924, an annual radio exhibition (the Funkausstellung, still held annually as the 'IFA'[14]) was held in Berlin, and in 1930 Albert Einstein spoke over the radio on the occasion of the exhibition's opening, bringing his voice to a vast audience.

Figure 2.1 22 August 1930: Einstein speaks on the radio to open the Radio Exhibition in Berlin

Whether or not Reichenbach was responsible for Cassirer's decision to lecture on the radio, having Einstein as a model no doubt added to his readiness to address the public this way. Cassirer spoke for 25 minutes, from 18.30 to 18.55 on two successive weeks on the radio station Deutschlandsender in its series of university lectures called 'Hochschulfunk'.[15] For this, Cassirer would have had to go to the Berlin suburb of Zeesen to the transmitter studio.

The Deutschlandsender carried its name for good reason: it could be heard throughout all of Germany and even in countries beyond its borders. If Reichenbach was behind the invitation to Cassirer to speak on the Deutschlandsender, then he might have been provoked by Cassirer's two-part lecture series, for although Cassirer's lectures were entitled 'The Unity of Science' ('Die Einheit der Wissenschaft'), Cassirer actually dealt with the plurality of ways in which humans understand the world, rejecting the underlying 'physicalism' that many adherents of the Logical Empiricism, such as Reichenbach, upheld. However, as Reichenbach's correspondence with Cassirer makes clear, both men held the other in high esteem, and their friendship continued over decades until Cassirer's death. Cassirer even supported Reichenbach in his

Figure 2.2   21 and 28 October 1931: Cassirer lectures on national radio on 'The Unity of Science' over the Deutschlandsender near Berlin in their broadcast 'Hochschulfunk'

attempts to have professors of philosophy appointed with special focus on the philosophy of science and logic.[16] Cassirer's radio lectures have only recently been published in the edition of his *Nachlass*.[17]

The changes wrought by the rise of the new technologies in the era of the Weimer Republic were so great that they became a frequent topic for artists, dramatists and performers. 'Technik' became the name, alternatively, for a threat to mankind's survival or the means for its redemption. Technik was treated in literature and drama as either a monster or as a saviour. Plays such as Georg Kaiser's *GAS*[18] (1918) or Max Brand's *Maschinist Hopkins* (1928),[19] a modern opera, were great successes, playing on stages throughout the country. In *GAS* Kaiser portrayed technology as a force leading to the destruction of society, while in *Maschinist Hopkins* technology was the locus for unified action and society's salvation from chaos. Fritz Lang's movie *Metropolis* (1927) brought the same sort of questions to the screen.

In any case, the sweeping changes in society resulting from 'complete mechanization' were obvious to all. The increased speed of life was new, and this too entered into popular culture and art. With the question whether technology was going to change life for the better or for worse uppermost in everyone's mind, it is not surprising that this is the topic that occupied Cassirer at the outset in his paper. Artists in particular could hardly have ignored these changes – and musicians especially, since the world was becoming a very loud place. The most popular opera of the 1920s was Ernst Krenek's *Jonny spielt auf* (1926),[20] which dealt among other things with the spread of jazz and which called in one scene for a locomotive to come onto the stage. The most dramatic film expression of mechanization was Walter Ruttmann's film *Berlin: Die Sinfonie der Großstadt* (1927) in which the city itself was the main actor – the footage was all shot on location without actors – and the main theme was restlessness.

Kestenberg's line-up of contributors to his essay volume was carefully chosen, and included essays by both Krenek and Ruttmann. But the book appeared at an inopportune time. In January of 1933 Hitler was named chancellor in Germany. Kestenberg, who was Jewish, had to flee the country, and this book was thereby no longer the sort of reference that could be cited. What is more, art and technology were matters of great importance to the new regime. The book had appeared in a limited edition, and circulated only among cognocenti. Only two years later the topics discussed in it were regarded in a new way. The National Socialist leaders in Germany made extensive use of radio to broadcast propaganda and to shape aesthetic sensibilities in a particular way. In 1946 Cassirer's last book, *The Myth of the State* focused upon what he called 'the technique of myth': the deliberate dissemination of manufactured mythic conceptions by modern technical means.[21] In this work he answered the question that had captivated the public and intellectuals in the 1920s – whether technology was destined to lead to the destruction of civilization – but differently than he was prepared to in 1930. His answer was that no such thing as 'destiny' was involved, but that technology could be used in ways that nobody had anticipated, to manipulate elementary human emotions, and, by turning fear into hatred, control large masses in ways that physical force could not. Instead of helping art to liberate human emotions, it expanded them so as to create a monstrous force that could be channelled by those in power to their own ends. The era of 'complete mechanization' ended with the mechanization of human beings themselves.

Notes

1. Aristotle, *Ethica Nicomachea*, X, 6–8.
2. Erwin Panofsky, *Idea: Ein Beitrag zur Begriffsgeschichte der älteren Kunsttheorie* (Studien der Bibliothek Warburg) (Leipzig and Berlin, 1924), 6 and note 22.
3. Plato, *Republic*, 596 D.
4. Originally published in German as 'Form und Technik' (editors' note).
5. See Ernst Cassirer's 'Form and Technology' in this volume, 54.
6. Siegfried Gideon, *Mechanization Takes Command: A Contribution to Anonymous History* (Oxford: Oxford University Press, 1948) (editors' note).
7. Leo Kestenberg's writings are to appear in a four-volume edition, edited by Wilfried Gruhn with Ulrich Mahlert, Dietmar Schenk and Judith Cohen. The two first volumes of this *Gesamtausgabe* appeared in 2010.
8. See Hans Curjel, *Experiment Krolloper, 1927–1931*, ed. Eigel Kruttge (Munich: Prestel, 1975).
9. Cassirer and Cohen: Histories, relatives and descendants. 'Kabelwerke Cassirer'. Accessed by the editors 26 September 2011, http://genealogy.metastudies.net/ZDocs/Cassirer/Cassirer_Falk_Miscellaneous/pages/29.html
10. The building survives today, protected as an example of modern architecture (Hugo-Cassirer-Straße 40); it houses the Berlin Stadtmuseum's archive, depot and restauration centre.
11. Poelzig was much sought-after in the 1920s, building, for example, Max Reinhardt's famous expressionist-styled theatre, the movie theatre 'Babylon', which still exists, and the sets for the expressionist film *Der Golem, wie er in die Welt kam* (directed by Carl Boese and Paul Wegener, 1920). See *Hans Poelzig: 1869 bis 1936; Architekt, Lehrer, Künstler* (Munich: DVA, 2007).
12. Hans Reichenbach, *Was ist Radio?*, 2nd edn (Berlin: R. C. Schmidt, 1929).
13. A radio station whose name means, literally, 'Germany transmitter' (editors' note).
14. 'Internationale Funkausstellung Berlin' ('International Radio Exhibition Berlin') is one of the oldest industrial exhibitions or trade fairs in Germany (editors' note).
15. 'University college radio' (editors' note).
16. See Cassirer's correspondence with Reichenbach in Ernst Cassirer, *Nachgelassene Manuskripte und Texte*, vol. 18: *Briefe: Ausgewählter wissenschaftlicher Briefwechsel*, ed. John Michael Krois (Hamburg: Meiner, 2008).
17. Cassirer, 'Die Einheit der Wissenschaft', in Ernst Cassirer, *Nachgelassene Manuskripte und Texte*, vol. 8: *Vorlesungen und Vorträge zu Philosophischen Problemen der Wissenschaften, 1907–1945*, ed. Jörg Fingerhut, Gerald Hartung and Rüdiger Kramme (Hamburg: Meiner 2008). Cassirer's first contact with radio broadcasting was not his last, for he lectured again in January of 1937 on René Descartes' *Discourse on Method* over Austrian radio. Cassirer had already left Germany in 1933, upon Hitler's rise to power, and was teaching in Sweden.
18. Georg Kaiser, *Gas: Schauspiel in fünf Akten* (Berlin: Fischer, 1918).

19. Max Brand, *Maschinist Hopkins: Oper in einem Vorspiel und drei Akten* (Wien and Leipzig: Universal Edition, 1928).

20. Ernst Krenek, *Jonny spielt auf: Oper in zwei Teilen* (Wien and New York: Universal Edition, 1926).

21. Ernst Cassirer, *The Myth of the State* (New Haven: Yale University Press, 1946). See Cassirer, 'The Myth of the State: Its Origin and Its Meaning. Third Part: The Myth of the Twentieth Century', in *Nachgelassene Manuskripte und Texte*, vol. 9: *Philosophie und Politik*, ed. John Michael Krois and Christian Möckel (Hamburg: Meiner, 2008), 167–224.

Part II
Contemporary Readings

3
Technics of Thinking

Aud Sissel Hoel

What is thinking? An ambitious question, to be sure. Yet this precise question was tackled by Ernst Cassirer in a highly suggestive and thought-provoking way in his essay 'Form and Technology' from 1930. This essay, delivered as a supplement to his three-volume *magnum opus* on the philosophy of symbolic forms (1923–29), sets out to determine the 'being' of technology. Cassirer poses the question concerning technology on the grounds that the philosophical depth and significance of this question has not been sufficiently acknowledged in the existing literature on the topic. So what, then, has technology to do with thinking? Viewed through the optics of the essay under discussion: everything.

Like knowledge, technology is ruled by a demand for some sort of 'correspondence' with reality. Cassirer is not content, however, only to draw an analogy between the two problems. Nor is he content to conceive technology as applied science. On the contrary, the pivotal point in 'Form and Technology' is that technology makes knowledge as such possible by grounding 'a type of mediacy that belongs to the essence of thought'.[1] The cognitive role ascribed to technology has far-reaching implications for what we are to make of the old requirement of agreement between thought and world. Indeed, it is exactly here, in Cassirer's radical reconception of what it means for something to 'correspond' with something else, that his approach breaks new ground and stands out at its most original. 'Truth' to Cassirer is not a dyadic relation between beliefs and facts, but a dynamic and, most notably, *differential* relation, which not only transforms and produces the object of knowledge but also – again worth noticing – *reveals* the object in and through this transformative production.

The argument to be made in the present chapter is that Cassirer, in his essay on technology, makes a significant contribution to what

65

could be conceived of as a *differential* approach to knowledge. As such, this position – if it were to be developed further – has the potential for cutting across the deadlocked disputes between realism (metaphysical objectivism) and anti-realism. Crucial to the new position, which I will designate as 'transformational realism', is Cassirer's positive recognition of human productive activity and intervention as essential factors in all kinds of knowledge gaining. Just as crucial is his insistence that knowledge is realized only in and through symbolic forms and material instruments. Thus, the line of thought to be pursued in this chapter attends to aspects of Cassirer's philosophy that advance *beyond* critical idealism and that, in so doing, *break* with the neo-Kantian framework that it is most commonly associated with.

The differential approach to be limned in this chapter also promises a fresh take on the problem of *mind*. The challenge offered by Cassirer's essay concerns the way that thought and technology are brought together in a new and inner constellation. This constellation initiates a trajectory towards a conception of the mind that is neither dualist nor physicalist. If we are to remain consistent with the differential line of thought, we have to reject the idea of a separate cognitive faculty that is assumed to exist and operate in perfect isolation from the body as a living biological system. Cognition is seen as embodied through and through. Nevertheless, and as Cassirer repeatedly and insistently reminds us, the human mind is characterized by the way that it continuously and incessantly outruns its natural basis. Human beings are symbol-making and tool-making animals. Human cognition, therefore, is neither natural nor cultural but irreducibly both. This is to say that cognition proper – what is commonly referred to as the 'mind' – is identified by Cassirer with the intellectual capacities that come into being only when a living, biological system is 'upgraded' by cultural interventions.

Even though Cassirer foregrounds dynamic processes of differentiation in most, perhaps all, of his systematic writings, he never uses the term 'differential' to designate his own work. Neither does he designate his own position as 'transformational realism'. Quite the contrary, he explicitly positions himself in the critical idealist tradition initiated by Immanuel Kant. It is my impression, however, that this outspoken identification with the Kantian tradition serves to hide Cassirer's most original insights – not only for the reading public but sometimes even for Cassirer himself. The way I see it, Cassirer's most innovative ideas come forth exactly at the points where his thinking *diverges* from the Kantian and neo-Kantian paths. Certainly, it is Kant who gives us the key to a dynamic and relational approach. But, as this chapter shows, the main lesson to be drawn from Cassirer's essay on technology is that *genuine relations are differential*, which is to say that they are mediated by 'foreign' material and historically constituted forms. Hence, it is only due to Cassirer's distinctive *take* on the Kantian insight, to the *changes* he introduces, that the transcendental approach turns differential. The functional definition of the concept alone, as developed by Cassirer in *Substance and Function* from 1910, does not suffice. It is only when the cognitive function is transformed into a *symbolic* function – as in the philosophy of symbolic forms developed in the 1920s – that a decisive step in a differential direction is made. Further steps are made in the 1930 essay on technology as well as in a body of posthumously published texts that were intended for a fourth volume on the metaphysics of symbolic forms.[2] The differential approach to knowledge to be outlined in this chapter was not fully developed as such by Cassirer himself. Yet it builds on Cassirer's differential insights and develops them even further.

Reframing the problem of technology

Contemporary culture is characterized by the way that technology infiltrates each and every area of human life. The *understanding* of technology, however, as Cassirer aptly observes, has not kept pace.[3] Surely, there is no shortage of voices that reproach or praise technology. Yet the majority of these assessments fail to provide insight into technology's 'essence'. The reason Cassirer gives for this deficiency is that most assessments limit their investigations to technology's *products* and *effects*. They limit themselves, in other words, to investigate technology as if it were a mere *thing*. Yet, as Cassirer points out, 'in the world of tools there are no mere things with properties'.[4] The difficulty of the problem has to do with the way that the 'being' of the tool makes itself manifest, not in its thing-like existence, but rather in 'the expression of a particular activity to be *performed*'.[5] If we are to come to grips with technology, therefore, we have to approach it from a different angle. We have to 'shift our gaze from the mere product to the mode and type of production and to the lawfulness revealed in it'.[6] All of this amounts to the same as saying that, if our investigations are to bear fruit, we have to stop thinking of 'essences' in the terms provided by substance metaphysics. The problem of technology needs an entirely new formulation; it has to be rephrased as a problem of *form*.

At this point the problem of technology touches on an overriding concern in Cassirer's thinking regarding the role of the functional concept in human knowledge. The core principle around which Cassirer's systematic philosophy revolves and that continues to spark his thinking throughout its various developments is *the primacy of the function over the object*. This principle as deployed by Cassirer has two sources of inspiration: first, the mathematical function in modern mathematics and, second, the 'Copernican turn' as performed by Kant's transcendental philosophy. In both cases the principle is taken to challenge the objectivist ontology that springs from substance metaphysics. In the present context I will focus mainly on the second source of inspiration.

On Cassirer's view, Kant's *Critique of Pure Reason* brought to philosophy nothing but a revolution in method. The revolutionary element had to do with the way that Kant gave the problem of the relation between cognition and its object, as generally accepted until then, an entirely new formulation. Cassirer sums up Kant's cardinal move as follows:

> Instead of starting from the object as the known and given, we must begin with the law of cognition, which alone is truly accessible and certain in a primary sense; instead of defining the universal qualities of *being*, like ontological metaphysics, we must, by an analysis of reason, ascertain the fundamental form of *judgment* and define it in all its numerous ramifications; only if this is done, can objectivity become conceivable.[7]

What particularly caught Cassirer's attention is the way that cognition is redefined by Kant as a productive *function* that embodies an original, formative power. Cognition is no longer seen as a copy of an absolute, self-subsisting reality; the object is, rather, seen as the *product* of cognition's structuring activity.[8] Kant's most important achievement, then, consisted in the establishment of a *genetic* view of knowledge. As Cassirer saw it, the *Critique of Pure Reason* laid down the direction for a new *critical* style of thinking that is characterized by the way it turns back upon itself and enquires into its own conditions. It is Cassirer's endorsement of this self-reflective style of thinking that warrants his explicit identification with Kantian idealism.

For all that, it is worth noticing that when Cassirer positions himself within the Kantian tradition he tends to put forth an interpretation that pulls Kant in a differential direction. In the posthumously published text, 'On Basis Phenomena' (written probably around 1940), for instance, where Cassirer makes a distinction between different kinds of theories of knowledge, he positions Kant together with his own philosophy of symbolic forms in the group of theories that are understood to start from what he calls the 'work-aspect'. By comparison, the phenomenological approach of Edmund Husserl is characterized by Cassirer as a 'monadic' theory that starts from the 'I-aspect', whereas the phenomenological approach of Martin Heidegger is positioned among the pragmatist theories that start from the 'action-aspect'.[9] The above classification could certainly be disputed. Instead of entering into a detailed discussion of Cassirer's proposed scheme, however, I will propose an alternative scheme, the purpose of which is to foreground the contrasts between a *differential* approach to knowledge drawn from Cassirer's differential insights and classical *transcendental* approaches such as Kant's and Husserl's (leaving Heidegger, for the time being, out of the equation).[10] For the sake of clarity, then, permit me to make some gross generalizations.

On my alternative scheme, the approaches of Kant, Husserl and Cassirer are first grouped together, since they are all critical theories of knowledge that endorse the principle of the primacy of the function over the object. They differ, however, in the way that they *interpret* this principle. Kant's radical move consisted in his replacing of the metaphysical model of subject and object with a relational model, where subject and object are understood, rather, as *correlates*. The Kantian model is dyadic (it consists of two elements) and bipolar, which means that the elements involved exist merely as dependent 'moments' in a mutual and reciprocal relation. The bipolar model is retained in Husserl, in and through the central role he ascribes to intentionality (all experience is experience *of* something). Besides that, the approaches of Kant and Husserl converge in the way they conceive the constitutive structures as *immanent*. This is why both thinkers privilege relations that are assumed to be *direct*, and why they assume the existence of something like pure cognition, or as in Husserl's case, pure consciousness. One of the problems with transcendental approaches such as these is that their main characteristics, as presented above, threaten to cancel each other out. Both Kant and Husserl give priority to the subject-pole of the correlation (hence the crucial role played by the transcendental subject in their respective approaches), with the result that the hard-won relational stance threatens to dissolve into monadic self-certainty.[11] Cassirer, from his side, also takes up the relational stance. But the philosophy of symbolic forms adds a twist that immediately sets him at a distance from the transcendental approaches: the constitutive structures are precisely *not* considered to be immanent. Quite the contrary, in Cassirer's account the constitutive structures are conceived to be material as well as historically constituted and as such they are essentially *foreign*. Furthermore, the constitutive structures are understood to realize the cognitive function, not in spite of their foreignness, but precisely by virtue of it. To put it differently: in Cassirer's account the realization of the cognitive function requires the intervention of a foreign parameter, of a 'terminus medius', which belongs neither to the subject-pole nor to the object-pole, but which serves instead as the differential principle that provokes the poles to diverge in the first place. This is what it means for a relation to be differential, and this is why cognition, in a differential account, is conceived as essentially mediated.

Finally, to tie this lengthy excursion back to the issue at hand: Cassirer's point in 'Form and Technology' is that in order for us to understand technology, we have to acknowledge that technology too embodies an original, formative power. We have to realize that technology is in fact one of the basic forms of culture, and as such it is liable to the same kind of critical self-reflection as the other forms. Just as much as we have to enquire into 'the "conditions of the possibility" of theoretical knowledge, language, and art', we have to enquire into 'the "conditions of the possibility" of technological efficacy and technological formation'.[12]

Expanding and transforming *logos*

In Kant's *Critique of Pure Reason*, mathematics occupies a privileged position to the extent of exhausting the field of theoretical knowledge. It should come as no surprise, therefore, that mathematics serves as the paradigm case in Cassirer's early investigations of the functional concept. Yet it is not until Cassirer realizes that the validity of the functional concept in fact applies to the entire field of human meaning-making that his thinking changes to a differential track. In Cassirer's account, 'the Copernican revolution with which Kant began ... refers no longer solely to the function of logical judgment but extends with equal justification and right to every trend and every principle by which the human spirit gives form to reality'.[13] The human intellectual domain is thus radically expanded, and *logos* is understood by Cassirer as a name for 'all those functions which constitute and build the world of human culture'.[14] Owing to this expansion, the cognitive function is *pluralized*: 'For the fundamental principle of critical thinking, the principle of the "primacy" of the function over the object, assumes in each special field a new form and demands a new and dependent explanation.'[15] Depending on the differential principle that is set in operation, cognition takes another direction. Cognition, therefore, is conceived by Cassirer as *multidirectional* and *multidimensional*.

Cassirer himself considered the expansion of the critique of reason to a critique of culture as a mere amplification of Kant's original move.[16] He seems to be oblivious to the way that the reformulation of the cognitive function into a symbolic function completely transforms the notion of *logos* as understood by Kant. In this respect it is symptomatic that Cassirer continues to privilege mathematics and to talk about 'pure thought', even after his philosophy has taken a symbolic turn. If we are to remain consistent with the differential line of thought, however, we have to realize that there is no such thing as 'pure cognition' and that mathematics, by implication, is but one of several productive constructive forms – a very sophisticated and highly developed form, for sure, yet these characteristics do not make mathematics the very form of 'cognition itself'.

A further step in a differential direction is made in the essay on technology, where Cassirer maintains that *logos* has an instrumental – not only theoretical – side to it. In this text he advances a *productive* view of knowledge by claiming that knowing and making, thinking and doing, are 'originally united' and stem from a 'common root'.[17] He substantiates this view by drawing a parallel between language and technology, taking his impetus from a lecture on poetry and technology by the German engineer and writer Max Eyth. According to Eyth, it is the word and the tool that distinguish the human being from the animal. Yet we fail to see the inner connection between the two if we think of language as the sole tool of the mind and of the tool as a mere means in the service of knowledge. Cassirer follows up on Eyth's argument by pointing to the instrumental side of language. We misconstrue the nature of language if we think of it only as a means for describing reality; it is, rather, a means for making reality. The intellectual nature of the human being is always double. The operations of language are, simultaneously, theoretical as well as instrumental. The thesis of the double nature of all intellectual activities also applies the other way around: 'implicitly contained in this is the counter-thesis that the potency of logos also resides in every simple material tool'.[18] Language and technology are both forms of *logos*, and as such they are both 'tools of the mind'.

Bios and *logos*

Cassirer's approach differs from classical theories of knowledge in the way that he understands the mediating role of the body. He rejects dualist models that posit the intellect, conceived as pure spontaneous action, against sensibility, conceived as mere receptivity. In Cassirer's view, universality and meaning do not begin with the concept. Even before the onset of symbols and tools, formative processes are in operation. There is, in other words, an 'activity of the sensibility itself'.[19] Instead of talking about passive impressions, therefore, Cassirer talks about a 'world of expressions'. The term 'expression' refers to a basic level of meaning that is assumed to be shared by all living, biological systems. It is introduced to accentuate the way that the world, for every living being, is always already meaningful. Expressive meaning is explicated by Cassirer in functional terms: Individual forms of life have access to different 'worlds' depending on their 'organization'. In this conception, Cassirer takes inspiration from the biologist Jakob von Uexküll and his definitions of the living organism as a 'functional circle' and of the world as an 'Umwelt'. At the basic level, then, meaning is 'physiognomic'. Cassirer explains:

> The 'surrounding world' [*Umwelt*] or environment of every animal cannot, in general, be defined by *our* concepts of 'objects' and 'characteristics'; rather, it stems from the whole of its organization and corresponds in every feature to this whole. In the world of the earthworm, there are only earthworm things; in the world of the dragonfly, only dragonfly things.[20]

The expression, then, is yet another instantiation of the principle of the primacy of function, but in this case it operates 'below' the level of objects. Compared to the 'higher' meaning functions, however, the expressive function is primary in that it provides the very opening onto the world that makes meaning possible in the first place. The living organism is understood as a 'circle of life' and as such it involves action: 'Whatever is alive has its own circle of action for which it is there and which is there "for" it – both as a wall that closes it off and as a "viewpoint" that it holds "open" for the world.'[21] Each individual form of life, then, amounts to a specific *mode* of having the world, a specific *way* of revealing and dealing with it. It is against this background, then, that we can understand Cassirer's characterization of 'life' and 'action' as 'basis phenomena'.

If we are to speak about knowledge, however, further mediations are necessitated than those provided by the body itself. For, as Cassirer maintains, it is only by virtue of the distance introduced by symbols and tools that the life complex becomes a knowledge complex.[22]

Human existence is a peculiar existence because it is *mediated in principle*. It is an existence, therefore, that *always already differs from itself*. While it never ceases to be a life complex, it constantly transcends its own natural foundations. There are several ways of interpreting this characteristic transcendence and, with that, the mediating role played by symbols and tools. One approach, which prevailed in Cassirer's time and which has continued to reassert itself in different guises ever since, understands the mediated character of human existence in terms of *alienation*. Ludwig Klages, for instance, who gives voice to this approach in Cassirer's essay, maintains that technology 'alienates the human being from his own centre of life', and, as a result of this progressive alienation, the human being has 'fallen out with the planet that bore and nurtured him'.[23] What characterizes Klages' approach and other approaches that foreground alienation is that 'life' and 'mind' are conceived as metaphysical forces that oppose and counteract one another – Klages' characterization of the forms of spirit as 'vampiric' and 'soul-destroying' powers being an obvious case in point.

Cassirer, on his side, sees things differently. The mediating role of symbols and tools is conceived by him, rather, in terms of *articulation*. Symbols and tools are not understood by Cassirer to alienate the human being from its 'true essence'. Quite the opposite, it is only in and through this intermediary that the human being is constituted as such and comes into itself as an 'I': 'The border that separates purely organic efficacy from this technological doing is likewise a sharp and clear demarcating line within the development of I-consciousness and singular "self-knowledge".'[24] Further, the intervention of cultural forms is not understood to 'close' the opening onto the world. They are understood, rather, to amplify the perception-action circuits of the life complex beyond themselves, making the organism able to perceive and do new things. For Cassirer, then, the duality of life and mind is a duality of viewpoints only. The distance introduced by the tool allows for a *new kind of seeing* and thus for a *new kind of relating* to the world. What is more, the distance introduced by the intermediaries opens up a space that allows the human being a certain *leeway* vis-à-vis its existential circumstances. The human being does not merely adjust to its niche; it actively builds it:

> And it is by moving about in this space, in the whole of that which is achieved through his work – and through which his work first becomes possible – that the human progressively builds up his world, his horizon of 'objects', and the concept of his own essence.[25]

This is why, in the case of the human being, there are not only two, but *three* basis phenomena to be factored in: life, action and *work*.

Where, then, does this leave us when it comes to the status of the human mind? Again, the answer will depend on whether or not we remain consistent with the differential trajectory of thinking (Cassirer's lines of thought are sometimes discordant at this point and waver between transcendental and differential conceptions). But if we do, and as Cassirer certainly does in his technology essay and in the posthumously published texts intended for the fourth volume, then 'mind' is seen as what comes into being only when a living, organic system is differentiated from itself through the formative intervention of cultural forms. Mind, therefore, far from being seen as life's irreconcilable opponent, is seen, rather, as a *transformation* – an upgrading even – *of life*.[26]

Truth, distance and intervention

Cassirer's approach to the problem of knowledge in 'Form and Technology' is radical for at least two reasons. First, by assigning constitutive powers and thus theoretical import to material tools, he violates the canonical distinction between theoretical, practical and technical modes of reasoning, which was set forth by Aristotle and which has informed Western thinking ever since.[27] Second, and perhaps contrary to what one would expect, this violation does *not* prompt Cassirer to renounce the idea of necessity or truth. Fully acknowledging the contingent and transitory nature of knowledge, then, he retains a notion of veracity.[28] As contradictory as this might sound, the point I want to make is that on a differential account, contingency and necessity do *not* rule each other out.

The notion of truth that results from a differential approach, however, is markedly different from the established notions in terms of coincidence or correspondence. In order to accentuate the difference, I will go back, if only very briefly, to antiquity. Before Plato and Aristotle limited the meaning of 'theoria' to the activity of contemplating eternal truths, the word referred to, as pointed out in a recent study, a religious practice that involved pilgrimage and witnessing of mysterious spectacles.[29] Interestingly, the connotation of sacred spectating has been kept in the philosophical use of the term. Aristotle, for instance, who defines 'theoria' as an intellectual activity that is cut off from the practical realm, understands theorizing activity as an actualization of the divine faculty in humans. In theorizing, the divine faculty becomes identical with the divine being, the primary substance, which is the first cause of all things.[30] From antiquity onwards, *pure intellectual presence to what is truly real* has continued to be taken as a standard of truth. Even in Kant, who insists on the finite nature of human reason, divine reason (aptly designated by him as 'intellectus archetypus') still figures as the standard of truth – even if here only as a negative standard that is unattainable for humans.

Since then, the spectator theory of knowledge has been subjected to numerous and incisive attacks, and rightfully so. What sets Cassirer apart from the main thrust of contemporary critics of 'ocular centrism' and 'the metaphysics of presence', however, is that he advances an approach that is not 'anti-essentialist'. For, as Cassirer sees it, going from one extreme to the other, say, from realism to relativism or scepticism, never solves the underlying problem. The taking up of a relativist or sceptic position, therefore, would not really challenge the essentialist model of truth; it would confirm it, rather, by providing its negative counter-image. In like manner, simply rejecting the notions of 'truth' and 'identity' on the grounds that the criteria set forth by the standard model are unattainable and impracticable would not really challenge the standard model; it would confirm it, rather, in a negative and roundabout way, by its continued adherence to the old criteria for what is to count as true. Instead of pursuing an anti-essentialist strategy, then, Cassirer chooses to approach the problem at a metalevel by challenging the very conceptual matrix that sustains essentialism as well as its negative reversals. He starts this work in *Substance and Function*, where he attacks the substantivist metaframework head on by challenging one of its main pillars, the Aristotelian notion of concept formation, and replacing it by a functional notion. The subsequent development of his thinking continues the enterprise of reframing knowledge and of establishing an alternative, and, as I see it, *differential* matrix or metaframework. Contrary to the standard model, which defines true knowledge in terms of passive beholding of pre-given essences, Cassirer holds that genuine knowledge is established through *work*, that is, through human ingenuity, invention and intervention. This is why he reminds us, in the essay on technology, that many great discoverers of nature were also inventors (Leonardo da Vinci serves as the paradigm case) and that 'theoria', 'praxis' and 'poiesis' interpenetrate each other – not only sometimes but in principle.

Cassirer's writings are brimful of visual metaphors. The symbolic forms amount to different 'ways of seeing', individual symbolic forms institute particular 'image worlds', concepts serve as 'viewpoints', tools initiate new 'lines of sight' and so forth. This way of putting things may cause dismay among contemporary thinkers who are, for very good reasons, wary of the privilege that has been accorded to vision throughout the history of Western philosophy. Yet it is precisely here, as regards Cassirer's notion of *visuality*, that the differential approach makes manifest, so to speak, its *difference*. For in Cassirer's account of the dynamics of meaning, the visual aspect is not understood in the traditional sense of direct perception of truth (intellectual contemplation), nor is it understood in the Kantian sense, as the heterogeneous counterpart of the 'discursive'. It is understood, rather, in terms of a creative *picturing* in accordance with some rule, much in the vein of the Kantian schemata. Yet, on the differential account, this creative intervention is not understood simply to impose its forms on experience. Instead, it is understood to *articulate* and *reveal* the phenomenon in and through the very act of exposing it. On a differential account, then, the discursive and the visual are two aspects of the very same process of differentiating. This is why the term 'viewpoint' may serve as a synonym for the term 'measure' in Cassirer's texts, and vice versa.

What is at stake here may become clearer if we relate the above discussion to the workings of tools. For what technology teaches us, before all else, is that human intervention does not preclude revelation. On the contrary, it teaches us that the 'discovery of nature' in fact *requires* the insertion of an intermediary. As Cassirer puts it in 'Form and Technology', the tool marks a 'turning point in *knowledge*' for the reason that it sets limits on the will. The tool introduces a distance that makes the will able to *look away from its goal*. The technical or mediated way of relating to the world manifests itself 'not only in the ability to seize its goal but also in the particular ability to distance the goal from it and to leave it at this distance, letting it stand there'.[31] The distance that is thus introduced is not merely negative; it does not solely involve a removal from the real. It is understood, rather, to bring the target phenomenon closer on a higher and more articulated level. The distance introduced by the tool is productive in that it allows us to discover the target phenomenon in its alterity, that is, as an *object* with an independent existence:

> The will can never succeed in its application simply by making itself stronger. Success demands that the will intervene in an originally foreign order and that it know and recognize this order as such. This knowing is at the same time a mode of recognition. Nature is not, as in magic, merely repressed by desiring and imagining. Rather, its own independent being is acknowledged.[32]

This 'discovery of nature' happens each time a technology is deployed, Cassirer maintains, no matter how simple the tool in question might be. He goes on from this to insist on technology's revealing capacity: 'This discovery is a disclosure; it is the grasping and the making one's own of an essential connection that previously lay hidden.'[33] It is only due to an active human intervention, therefore, that reality is provoked to reveal something about itself. Yet, even if these revelations are effected by human intervention, there is nothing arbitrary about them: 'There is no uncertainty, no mere subjective insecurity attached to these possibilities; they confront thought as something thoroughly objective.'[34] This, then, is how the differential approach brings necessity and contingency together: the revelations follow by necessity from the 'viewpoint' that is introduced by the merely contingent tool. This implies that another tool would reveal the target phenomenon differently. So, even though there is nothing arbitrary in the way that a tool reveals reality, the revelations in question will always remain relative to the viewpoint of the tool that exposed them. On a differential account, therefore, there is no such thing as a unique and exhaustive description of a phenomenon. All phenomena can be determined in multiple ways. For, according to Cassirer, the human being 'does not already *have* a determined form of the world; he must instead *search* for this form and *find* it in various ways. The way he finds it depends on the dynamic principle that the general movement of mind follows.'[35] This, then, is why Cassirer is able to talk about 'relative truths' without contradicting himself.

Considering all of the above, Cassirer's dismissive attitude towards pragmatist philosophy seems rather out of place. Surely, Cassirer disapproves of the utilitarian motives that sometimes surface in pragmatist thinking. Aristotle considered theoretical knowledge to be disinterested and nonutilitarian,[36] and even if Cassirer considers all knowledge to be productive, he was certainly not willing to go to the opposite extreme by 'defining theoretical "truth" merely as a special case of "utility"'.[37] Aside from that, there are obvious overlaps with pragmatism of which Cassirer seems to be oblivious. There may be several reasons for this, philosophical as well as circumstantial. Suffice it to say that Cassirer associates Heidegger with pragmatist philosophy, and that some of his misgivings come up during the famous disputation between the two thinkers in Davos, Switzerland, in 1929. Both thinkers defend their respective interpretations of Kant, and the discussion turns on whether finite human beings are able to 'break through' to a transcendental sphere and attain necessary and eternal truths. Heidegger targets the philosophy of Hermann Cohen, Cassirer's teacher, and addresses Cassirer as a neo-Kantian thinker. Cassirer, on his side, confirms this set-up by identifying his own philosophy with Cohen's.[38] With one blow, then, the distance Cassirer has travelled beyond neo-Kantianism is abolished. As a result, the two thinkers position themselves at the maximum distance from each other: Heidegger resolutely asserting human finitude and Cassirer invoking objectively necessary truths in mathematics. However, this set-up underplays the fact that the two thinkers converge, at least partly, on the question of the nature of human transcendence.

In Davos, both Heidegger and Cassirer understood transcendence as something that arises out of productive imagination. Yet, as Heidegger reminds us, this is tantamount to saying that transcendence remains *within* the sphere of the humanly produced, which again is an argument for its finite character. Cassirer, in replying to a question from the audience regarding the possible routes to transcendence, clarifies that transcendence can only be obtained through the medium of *form*, and for this reason the resulting transcendence will always remain an *immanent* transcendence.[39] In this way both thinkers accentuate, as paradoxical as it might sound, that transcendence is possible only *within* a human horizon. The paradox dissolves, however, with a view to the role the two thinkers attribute to 'poiesis', and not only to 'praxis' (Heidegger perhaps more so in his later writings). In their own idiosyncratic ways, both thinkers seem to gesture towards what I will designate as a 'poetic' notion of infinity. I will elaborate on what I mean by this notion by building on some of Cassirer's scattered references on the topic.

Poetic infinity

It is conceivable that Cassirer's grievance against pragmatism had to do with what he saw as an overemphasis on the 'action-aspect' at the expense of the 'work-aspect'. For what characterizes human action is that it tends to *settle*. It does not remain in a process of pure becoming. I cite from the text on basis phenomena, which comments on the difference between these two aspects and how they relate to each other:

> The 'work' appears, in contrast to the level of 'action' ... as something objective and, to an extent, fixed. The work is the aim of 'action'; but in this action it also comes to its end ... The movement of action has come to a halt; it has found expression in a work.[40]

This settling-of-action-into-work is, of course, the founding moment of culture. Human action tends to sediment, and in and through this sedimentation it obtains a kind of endurance. The crucial characteristic of the work has precisely to do with this peculiar 'temporal shape' and not merely with its 'technical usefulness' or its 'effects on the souls of men':

It is this 'being' that 'outlives' the moments which are not dragged into the turmoil of physical and psychical activity as it changes from moment to moment – this is the basic determining factor in the make-up of a 'work'.[41]

Furthermore, 'endurance' is the hallmark of the poetic and exactly what distinguishes it from the merely practical:

The practical is directed toward an effect in the present, as something momentary, toward an 'influence' on physical nature or on the human will. The poetic is different because its being is not limited *only* to such works. The poetic 'arises' and 'endures' outside every 'intention' (as 'aiming at a goal' taken as a specific, momentary, individual action). It is 'without interest'.[42]

In the last quotation, Cassirer brings up yet another characteristic that is understood to distinguish the poetic from the practical: the poetic is *disinterested*. Again, Aristotle's distinctions come to mind, but this time, of course, Cassirer alludes to Kant's treatment of aesthetics in the *Critique of Judgment*. In 'Form and Technology', art is brought in as a third basic form of culture, along with the word and the tool. The context is a discussion of Friedrich Schiller's notion of the 'ludic drive' ('Spieltrieb') and his idea that art is a 'creator of humanity' in that it 'makes possible the specific "mode" of being human'.[43] For what distinguishes the human being, according to Schiller's *Aesthetic Letters*, is its capacity for 'free play'.[44] And here we have arrived at the point where Cassirer makes one of his most provocative moves: he suggests that the free play of the imagination is *not* limited to the sphere of art or to the domain of the aesthetic, as understood, for instance, by Schiller and Kant. For, as seen above, the tool is also a 'creator of humanity', and in that respect perhaps even more fundamental. I quote once more from 'On Basis Phenomena': 'The transition to the "enduring" work (product) and to the tool as something which is "always to be applied in the same way" is what actually opens up to mankind the "objective" sphere, the sphere of "things".'[45] In 'Form and Technology', he also comments on the 'transcendence' of the work tool: 'The tool signifies something detached from its immediate being and becomes something that exists in itself, a continued existence that can far outlast the life of the individual human being.'[46] And again, in yet another context, Cassirer accentuates the way that the poetic extends far beyond the sphere of art, since 'absence of interest is not confined to the work of art alone, but holds equally for works of language, philosophical works, works of science, and pure knowledge in general'.[47] Innocent as they may sound, these quotations actually break new ground: the area of the poetic is expanded so that it comprises not only the 'fictional' but the 'factual' as well. Taken together, these quotations are suggestive of a fresh and highly original take on the problem of knowledge, which revolves around the insight that *objectivity is poetic*. Even so – and here comes the differential twist – the poetic determination of objectivity *does not rid the factual of its necessity*.

Technology and the possible

The tool, then, is the key to transcendence because it introduces a distance that makes humans able to see and act beyond the situation here and now. Or, to put it differently, the tool institutes and opens the sphere of the *possible*. The tool makes possible a new line of sight that has to do with the way that it installs itself between the will and the sought-after goal, and, due to this in-between position, pushes the goal into the distance. 'Instead of looking spellbound at this goal, the human being learns to "fore-see" it.'[48] The tool marks the onset of a capacity for forethought and, with that, a capacity for considering different ways of attaining it. For the tool-making animal, then, 'reality shows itself, regardless of its strict and irrevocable order, not as an essentially rigid existence but rather as a modifiable, malleable material. Its gestalt is not complete. Rather, it offers human will and initiative an enormous latitude for action.'[49] It is this differential reconfiguration of visuality and discursivity, of necessity and contingency, that makes it possible to rethink the old problem of 'correspondence' with the world.

As Cassirer points out, the 'battle of the systems' in philosophy turns precisely on the problem of correspondence. Whether they hold that knowledge is a *part* of being or whether they hold that being *effects* thought, all dogmatic systems fall through because they begin with the notion of substance.[50] What approaches such as these all fail to see is that the knowledge requires 'the use of an active force'.[51] This is why Cassirer, in the essay on technology, insists that theoretical knowledge and technical work are processes that touch upon each other. For in neither case is correspondence simply given; it has to be 'searched for and continuously produced'.[52] Thus understood, knowledge has no absolute source, neither in the subject nor in the object. It consists, rather, in an interminable process of mutual determination, provoked by the insertion of a '*terminus medius*' that serves as a differential viewpoint or measure: 'Mind always measures anew objects in relation to itself, and itself in relation to objects, in order to find and guarantee in this twofold act the genuine *adaequatio*, the actual "appropriateness", of both.'[53] It is worth noticing that Cassirer conceives of the determination process in terms of both *reduction* and *growth*: reality is reduced 'to its inner relation of measure' but looses nothing of its 'plasticity'; it is set, rather, 'as if in a fixed intellectual framework limited by certain rules of the "possible"'.[54] It is exactly this inscription into a larger intellectual framework, into an *order of possible determinations instituted by the insertion of a differential viewpoint*, that makes the phenomenon grow beyond itself. Cassirer adds the following specification:

> This inner growth does not simply take place under a continuous leadership, under the rule and guardianship of the actual; rather, it demands that we constantly return from the 'actual' to a realm of the 'possible', and see the actual itself according to this image of the possible.[55]

Finally, then, we have arrived at what Cassirer considers to be the greatest achievement of technology: 'Standing in the middle of the sphere of necessity and remaining within the idea of necessity, it discovers a sphere of free possibilities.'[56] Thus, from a theoretical point of view, technology's greatest achievement is that it demonstrates, in practice, that the notion of 'reality' as traditionally understood has been far too narrowly conceived. For as Cassirer observes, 'the sphere of the "objective" ... never coincides with the sphere of that which is presently at hand'.[57]

A differential metaframework, therefore, leaves us with a notion of truth that is worldly and relative but never 'relativist'. The human capacity for transcendence is understood neither in terms of an exertion of an alleged divine faculty in humans nor in terms of a taking up a 'perspectiveless' perspective; it is understood, rather, in terms of an active insertion of a differential sight, of an optic device that in a more than metaphorical sense aids the eye in its aiming.

Science criticism

Cassirer's essay 'Form and Technology' calls for a new kind of technology critique that does not limit itself to simply reproaching or praising technology's effects. This is very much in line with the concerns of the contemporary philosophy of technology as well as with the concerns of certain strands of science and technology studies (STS). These current approaches proceed based on the idea that technological artefacts are essentially non-neutral. The French sociologist Bruno Latour, for instance, who is one of the major proponents of STS, understands science in terms of actor-networks, that is, in terms of networks of relations that include both human and non-human agents. Technological agency also forms the keystone of an approach in current philosophy of technology called postphenomenology, which has been initiated by the philosopher Don Ihde. Like Cassirer, Ihde also includes reflections on technology's role when it comes to understanding human experience and the human existential condition.

As Cassirer observes, the thinkers who have put the most effort into understanding the effects of technology are typically the very same thinkers who have advanced a damning judgment over it.[58] Yet, as he points out in the essay, if we really want to understand a phenomenon, unmitigated condemnation is not very helpful. Postphenomenology is based on a similar conviction, which explains why today the focus has shifted from alienation, which prevailed in the classical philosophy of technology, to *technological mediation*. Like Cassirer, postphenomenological thinkers consider material media to be at the heart of human meaning formation and of human knowledge production. They proceed, in other words, based on a positive notion of technology's transformative role, something that differs markedly from classical philosophers of technology (as well as classical phenomenologists), who tend to privilege immediate experience over mediate experience. The new emphasis on technological mediation also resonates well with Cassirer's criticism of the nostalgic motive that underpins the alienation argument. Approaches that accentuate technology's alienating effects tend to conceive experience in terms of 'life', in terms of an unbroken continuity between the organism and its immediate surroundings. Technology, then, is understood to *break* this unity, to *interrupt* the alleged paradise of immediate existence. Cassirer, of course, renounces this utopian idea of a lost unity, and considers it futile to compare the present situation 'with the idyll of some pre-technological "state of nature"'.[59] The nostalgic motive also reappears *within* the field of technology, where it gives rise to a peculiar hierarchy between different kinds of technologies: the simple tool, which is assumed to form an unbroken continuation of the craftsman's natural movements, is privileged over the advanced tool, which 'breaks into the free rhythm of natural movements with a foreign dimension and foreign norm'.[60] Cassirer goes on to reject this hierarchical distinction, and he does so for two reasons. First, for him, *all* technologies, no matter how simple, introduce a foreign norm; indeed, the introduction of a foreign norm is the very definition of what technologies do. Next, he refuses the negative evaluation of 'foreignness' that is implied by the alienation argument. For, as we have seen, it is only by virtue of inserting a foreign parameter that the tool can fulfil its function as a 'tool of the mind'.

There is yet another reason to question the nostalgic motive that has proved so tenacious, especially in what the philosopher Carl Mitcham refers to as the 'humanities philosophy of technology'.[61] For, if 'true' existence is associated with immediate lifeworld transactions, contemporary science, with its abstractive methods and heavy instrumentation, may end up being seen as a reductionist enterprise almost by definition. And certainly, there are still, even today, philosophical voices that take a dismissive attitude towards science, especially, perhaps, among thinkers who emphasize the existential dimension of human experience. Yet again, if we want to *understand* science, sweeping dismissals are not very productive. This is why Ihde, much in the spirit of Cassirer, calls for a different *kind* of science criticism where the irrelevant humanist is turned into an authoritative critic.[62] What Ihde pleads for, to be more specific, is a humanist critic who enters the research phase of technological innovation and who participates in the shaping of scientific experiments. Participation in ongoing processes of technology development is also considered crucial by current advocates of an approach called constructive technology assessment (CTA), the overall aim of which is to contribute to 'the realization of better technologies (in a better society)'.[63] The proposed strategy of the CTA approach is to modulate the design process of new technologies by allowing broader negotiations (discussions, analyses, experiments) involving other concerned parties during the actual technology construction process. The purpose of these negotiations would be to anticipate the societal impacts of new technologies and to feed these insights back into the design process. The concerns of the CTA approach, then, resonate well with Cassirer's comments towards the end of his essay on technology, where he maintains that even if our society faces major technological challenges, technology alone cannot solve them. This is why Cassirer – and contemporary thinkers such as Latour, Ihde, CTA proponents and others – call for reflexivity and for societal negotiations – that is, for a new kind of *sociotechnical criticism*.

Extended mind

When it comes to understanding human cognition, the groundbreaking element in Cassirer's essay on technology has to do with the way that it offers an approach to cognition that is neither 'dualist' nor 'physicalist'. Dualist approaches such as René Descartes', which see mind and body as distinct substances, have long since fallen into disrepute. The contemporary study of mind is dominated, rather, by a multifarious collection of physicalist and functionalist[64] approaches. Physicalism is a philosophical position which holds that everything that exists can be explained exhaustively in terms of its physical attributes, effectively reducing questions of mind to questions of the material brain. Physicalist theories of the mind are clearly at odds with Cassirer, who, as we have seen, holds that meaning, even at its most primary and primitive level, constantly outruns the physical in the narrow sense that is usually employed by physicalists (the concept of the physical is limited to whatever can be described in physics). Moreover, in the case of human beings, Cassirer is wary of putting too much emphasis on the role of the body, due to the risk of overlooking the work-aspect and thus of reducing human meaning to a mere function of the body's physiological make-up:

> The concept of mankind is defined for it not by any specific identifiable structural features, but through the comprehensive totality of mankind's achievements. The totality of these achievements can in no way simply be read off from mankind's 'organization', such as from the organization of the brain and the nervous system.[65]

Functionalist theories of the mind differ from physicalist theories in that they conceive of mental states solely in terms of the causal roles they play in the cognitive system irrespective of how these states are materially implemented (in a biological brain, in a computer, or in some other system). The abstract and mechanist account of the cognitive process provided by functionalist theories is discordant with Cassirer's ideas of expression as a phenomenon of life and of mind as a transformation of life. Besides, since functionalism is based on an analogy between cognitive operations and the workings of mechanical devices (rendering the terms cognition and computation practically interchangeable), functional approaches cannot really account for the productive *difference* that is introduced when a biological system is hooked up with some 'external' device to form a coupled system. Just like the physicalists, then, most functionalists miss out on the work-aspect and, with that, the constitutive, augmentative and transformative power of symbols and tools.

For all that, within the contemporary cognitive sciences there *are* thinkers who *do* take the transformative power of contingent and 'external' resources into account. The philosopher Andy Clark, for instance, argues that the system we commonly refer to as 'mind' in fact extends beyond the resources provided by the biological brain. Taking inspiration from research fields such as animate vision and real-world robotics, which ascribe cognitive roles to bodily actions as well as to features of the environment, Clark advances an approach that is commonly referred to as the 'extended mind thesis'. From this perspective, cognition is understood to be 'scaffolded', that is, to rely partly on 'external' (bodily and environmental) features. Cognition, therefore, is understood to form a complex and coupled brain–body–world system. As Clark sees it, the classical view of cognition in terms of internal representations needs to be supplemented by an account of how some biological systems are able to exploit the environment, not only for the sake of 'minimizing computational effort', as he puts it, but also for the sake of augmentation. Scaffoldings allow the cognitive system to achieve goals that would otherwise be beyond it. And when it comes to human beings, language is understood by Clark to form a particularly powerful augmentative device.[66] These ideas clearly resonate with Cassirer, and in hindsight one could, perhaps, say that what Cassirer actually did, not only in 'Form and Technology' but also in his general approach to symbolic forms, was to advance an 'extended mind approach' *avant la lettre*. Indeed, when it comes to ascribing cognitive roles to external ('foreign') factors, he was even more radical than Clark, since to Cassirer human cognition is *essentially* scaffolded – not only occasionally so, as Clark seems to suggest. This is to say that, in Cassirer's account, the extended mind thesis does not merely supplement the internalist view of the mind but in fact replaces that approach. For, if we are to remain consistent with the differential approach to cognition, the bracketing out of the living body and of the external cognitive props (symbols and tools) would result in a bracketing out of the very phenomenon that we purport to study, that is, the very mind itself.

Transformational realism

As we have seen, according to Cassirer, the only transcendence that is available to humans is an *immanent* transcendence – a transcendence, that is, which in some sense is created by human beings (hence the term 'poetic infinity'). This prominence accorded to human productive forces in the process of knowledge gaining may prompt us to raise the old problem of *realism*. To what extent do we know that the world that is described in science is in fact the real world? To what extent do we know that reality exists independently of human observers and that it is not merely a projection of the human mind? This way of posing the problem, however, which is the traditional way, tacitly presupposes the substantivist metaframework that Cassirer sets out to challenge. To follow up on that, the reason why Cassirer repeatedly invokes the name of Kant, despite their obvious differences, is that he finds in *Critique of Pure Reason* the germ for a new metaframework. Additionally, if we take into consideration the general impact Kant has had on philosophy, we could say that since his time, the 'battle of the systems' has turned on how one is to interpret his cardinal insight regarding the limited nature of human understanding, and how one is to deal with its consequences.

It is not generally acknowledged that the *Critique of Pure Reason* challenges the very metaframework of Western philosophy, as Cassirer seems to think. On the contrary, to a considerable extent the presuppositions and conceptual grids delivered by substance metaphysics still serve as the default metaframework that is tacitly taken for granted. Viewed from *within* a substantivist metaframework, then, the Kantian insight concerning human finitude is typically interpreted in negative terms as an earnest reminder about the insufficiencies of the human intellect compared to the perfect intellect of, say, a godlike being who would be able to know things as they 'really are' and not only as they 'appear' to a human observer. The physicist and philosopher of science Abner Shimony, for example, connects the Kantian insight with a loss of naivety. According to Shimony, contemporary philosophers of science are now ready to admit 'the pervasiveness of subjective contributions to human representations', as he puts it.[67] For all that, Shimony is not willing to accept what he takes to be the Kantian idea that human knowledge is limited, in principle, to mere appearances. Nor is he willing to accept the philosopher of mind Hilary Putnam's contention of reality being, in principle, language- or mind-dependent.[68] Taking his inspiration from Aristotle, Shimony's project is to integrate epistemology with metaphysics and the natural sciences. This means that the knowing subject is to be understood as an entity in nature – that is, as an entity whose cognitive powers can be explained exhaustively from within a naturalistic worldview that emphasizes causal relations between mental and physical events. What is of relevance here, to the concerns of the present chapter, is how this project of 'naturalizing' human cognition moves Shimony to view all kinds of human formative powers with suspicion as potential sources of error. He quotes approvingly from Francis Bacon when the latter warns against false notions or 'idols' that may divert the cognizing subject from knowing the truth of nature:

For it is a false assertion that the sense of man is the measure of things. On the contrary, all perceptions as well of the sense as of the mind are according to the measure of the individual and not according to the measure of the universe. And the human understanding is like a false mirror, which, receiving rays irregularly, distorts and discolours the nature of things, by mingling its own nature with it.[69]

As should be clear from this, Shimony advocates a scientific realism that adheres to a substantivist standard of truth, where the scientific object is conceived as a self-subsisting entity. He considers it a sound strategy, therefore, to try to distinguish the 'intrinsic' properties of the object from the qualities that are merely 'projected onto it' by human beings. The formative powers that are involved in perception and cognition are understood by Shimony to cause distortions. But in an attempt to vindicate, as he says, the optimist spirit of Bacon, he believes it possible to *correct* these 'generic imperfections' of the human cognizing system. If we fine-tune our methods, if we study the cognitive apparatus *scientifically* – that is, through neurophysiology, perceptual psychology, cognitive psychology and related disciplines – *then* it will be possible to identify and correct all kinds of subjective sources of error, including those fallacies that originate in language or culture.[70] As should be clear from this, Shimony invariably treats human formative powers in negative terms, as they are associated with false notions, distortions and imperfections. And when it comes to the philosophical problems of language and culture, he effectively reduces these to questions of *constraints* that are *imposed on* the proper operations of the mind, which is conceived by him as an essentially natural system.

What I find to be Cassirer's main contribution to epistemology has to do with the way that he, in contrast to scientific realists such as Shimony as well as to relativists and sceptics, sets out to develop a *positive* notion of the 'limits' of cognition. It is with a view to this audacious undertaking that he advances one of his very few explicit criticisms of Kant: even Kant, Cassirer points out, seems to focus more on the negative consequences of his doctrine – that humans cannot know 'things in themselves' – than on his new positive insight – namely, that the genuine objectivity of knowledge is based on and secured in the free spontaneity of human beings.[71] Following up on this, Cassirer reworks the notion of knowledge so that it comes to involve an active 'grasping'. In 'Form and Technology' he maintains that theoretical thought has an instrumental side to it. It comprehends reality by '"gripping onto it" through the medium of *efficacy*'.[72] With his active and interventional notion of knowledge, and with his idea of the crucial role played by material tools in cognizing, Cassirer comes close to a diverse group of contemporary philosophers of science whom Don Ihde brings together under the term 'instrumental realists'. This group of thinkers, which apart from Ihde himself includes names such as Hubert Dreyfus and Ian Hacking,[73] are all critical of the classical philosophy of science with its abstract approach to knowledge and its penchant for 'rational' (linguistic, logical, or propositional) methods. The instrumental realists emphasize instead science as *practice*, and particularly, the embodiment of science in *technology*. Scientists are understood by these pragmatic technology-oriented philosophers to intervene into reality through instruments that enhance perception and make phenomena manipulatable. The instrumental realists diverge, however, on the question of whether or not scientific instruments 'deliver' reality. Speaking on his own behalf, Ihde sounds almost like Cassirer when he maintains that scientific instruments *both* transform *and* deliver the target phenomenon. The same is the case when he characterizes the procedure of measuring as 'a way of seeing' and science as a 'specialized mode of world exhibition'. However, whereas Ihde sees this shift from 'theory-prone' to 'praxis-prone' approaches as an indication of a general *shrinking* in size and significance of 'the territories previously taken as theoretical', leaving us with a notion of theorizing as 'a highly speculative exercise of scientific imagination',[74] Cassirer opts instead for an interpretation in terms of an *expansion* and subsequent *transformation* of the theoretical domain as traditionally understood. For, as we have seen, granting formative or transformative capacities to technology is for Cassirer tantamount to granting it a theoretical or cognitive import. The domain of *logos*, therefore, extends to material tools and instruments – including the procedures of measurement and the experimental techniques of science. And again the extension applies in both directions. For on a differential account, theorizing – no matter how 'speculative' – is never pure. The classical philosophers of science, therefore, fell short, not only because they ignored the cognitive import of scientific instruments, but also because they failed to acknowledge the embodied and material nature of their own rational methods. For, as Cassirer makes clear in

'Form and Technology', symbols too have an instrumental side to them; symbols too are 'tools of the mind'.

Notes

1. Ernst Cassirer's 'Form and Technology' in this volume, 30.
2. Ernst Cassirer, *The Philosophy of Symbolic Forms*, vol. 4: *The Metaphysics of Symbolic Forms*, trans. John Michael Krois, eds John Michael Krois and Donald Phillip Verene (New Haven: Yale University Press, 1996).
3. Cassirer, 'Form and Technology', 17.
4. *Ibid.*, 32.
5. *Ibid.*
6. *Ibid.*, 20.
7. Ernst Cassirer, *The Philosophy of Symbolic Forms*, vol. 1: *Language*, trans. Ralph Manheim (New Haven: Yale University Press, 1953), 78, original emphasis.
8. Cassirer, *Philosophy of Symbolic Forms*, vol. 1, 78. See also Ernst Cassirer, *The Philosophy of Symbolic Forms*, vol. 3: *The Phenomenology of Knowledge* (New Haven: Yale University Press, 1957), 315.
9. Cassirer, *Philosophy of Symbolic Forms*, vol. 4: 166–90.
10. The 'Kant' referred to is the Kant of the *Critique of Pure Reason* (first published in German in 1781/87), and the 'Husserl' referred to is the Husserl who has just made a 'transcendental turn' in *Ideas Pertaining to a Pure Phenomenology and to a Phenomenological Philosophy. Book 1: General Introduction to a Pure Phenomenology* (first published in German in 1913).
11. This interpretation of Kant and Husserl is disputed. In the reception of Husserl, especially, there is much discussion concerning the status of the 'noemata': Husserl distinguished between two correlated aspects of intentionality, 'noesis' which refers to the very process of cogitation and 'noema' which refers to that which this process is about (what is cogitated). The discussion centres on whether 'noema' should be understood as an ideal meaning (a conceptual or propositional representation) or whether it should be understood as the intended object itself as appearing. See Dan Zahavi, *Husserl's Phenomenology* (Stanford, CA: Stanford University Press, 2002), chapter 2.
12. Cassirer, 'Form and Technology', 18.
13. Cassirer, *Philosophy of Symbolic Forms*, vol. 1, 79.
14. Ernst Cassirer, 'Structuralism in Modern Linguistics', *Word: Journal of the Linguistic Circle of New York* 1 (1945), 99–120, 114.
15. Cassirer, *Philosophy of Symbolic Forms*, vol. 1, 79.
16. *Ibid.*, 79.
17. Cassirer, 'Form and Technology', 24.
18. *Ibid.*, 23.
19. Cassirer, *Philosophy of Symbolic Forms*, vol. 1, 87.
20. Cassirer, *Philosophy of Symbolic Forms*, vol. 4, 43.
21. *Ibid.*, 212.
22. *Ibid.*, 213.
23. Klages paraphrased by Cassirer, 'Form and Technology', 35.
24. Cassirer, 'Form and Technology', 37.

25. *Ibid.*, 34.
26. As pointed out by John Michael Krois and Donald Phillip Verene in their introduction to Cassirer, *Philosophy of Symbolic Forms*, vol. 4, xviii. See also Cassirer, *Philosophy of Symbolic Forms*, vol. 4: 213: 'All culture takes place in and proves itself in the creative process, in the activity of the symbolic forms, and through these forms life awakens to self-conscious life, and becomes mind.'
27. See Andrea Wilson Nightingale, *Spectacles of Truth in Classical Greek Philosophy* (Cambridge, MA: Cambridge University Press, 2004), 38.
28. Christiane Schmitz-Rigal, 'Ernst Cassirer: Open Constitution by Functional A Priori and Symbolical Structuring', in Michel Bitbol, Pierre Kerszberg and Jean Petitot (eds), *Constituting Objectivity: Transcendental Perspectives on Modern Physics* (Dordrecht: Springer, 2009).
29. Andrea Wilson Nightingale gives a detailed account of the cultural context of Greek philosophy, including 'theoria' as a cultural practice. In ancient Greece a 'theoros' was a pilgrim and it is not settled whether the word 'theoria' derives from 'theos' (god) or 'thea' (spectacle, sight). Nightingale, *Spectacles of Truth*, 40 and 45.
30. Nightingale, *Spectacles of Truth*, 5–6 and 13–14.
31. Cassirer, 'Form and Technology', 29.
32. *Ibid.*
33. *Ibid.*, 30.
34. *Ibid.*, 44.
35. *Ibid.*, 26.
36. Nightingale, *Spectacles of Truth*, 38.
37. Cassirer, 'Form and Technology', 16.
38. See the English translation of the Davos encounter record, Carl H. Hamburg, 'A Cassirer-Heidegger Seminar', *Philosophy and Phenomenology Research* 2 (1964), 208–22, 214.
39. Hamburg, 'A Cassirer-Heidegger Seminar', 215–18.
40. Cassirer, *Philosophy of Symbolic Forms*, vol. 4, 141.
41. *Ibid.*, 183.
42. *Ibid.*, 183.
43. Friedrich Schiller paraphrased by Cassirer, 'Form and Technology', 36.
44. Friedrich Schiller, *Über die ästhetische Erziehung des Menschen in einer Reihe von Briefen* (first published 1794). See especially letters 14, 15 and 21.
45. Cassirer, *Philosophy of Symbolic Forms*, vol. 4, 142.
46. Cassirer, 'Form and Technology', 32.
47. Cassirer, *Philosophy of Symbolic Forms*, vol. 4, 183.
48. Cassirer, 'Form and Technology', 31.
49. *Ibid.*, 34.
50. Cassirer, *Philosophy of Symbolic Forms*, vol. 4, 195–6.
51. Cassirer, 'Form and Technology', 43.
52. *Ibid.*
53. *Ibid.*, 44.
54. *Ibid.*, 30.
55. *Ibid.*, 44.
56. *Ibid.*
57. *Ibid.*

58. *Ibid.*, 47–8.
59. *Ibid.*, 41.
60. *Ibid.*, 40.
61. Carl Mitcham, *Thinking through Technology: The Path between Engineering and Philosophy* (Chicago: University of Chicago Press, 1994).
62. Evan Selinger, 'Introduction', in Evan Selinger (ed.), *Postphenomenology: A Critical Companion to Ihde* (Albany: State University of New York Press, 2006), 9; Don Ihde, *Expanding Hermeneutics: Visualism in Science* (Evanston, IL: Northwestern University Press, 1998).
63. Johan Schot and Arie Rip, 'The Past and Future of Constructive Technology Assessment', *Technological Forecasting and Social Change* 54, (1996), 263.
64. Not to be confused with Cassirer's 'functional' approach. In this passage the term 'Functionalism' refers to a theory of mind that explains mental states in terms of the functional or causal role they play in the cognitive system. Functionalism focuses solely on the effective functions of the brain, which are conceived in analogy with software programs. Influential proponents are Hilary Putnam, David Marr and Jerry Fodor.
65. Cassirer, *Philosophy of Symbolic Forms*, vol. 4, 43.
66. Andy Clark, *Being There: Putting Brain, Body, and World Together Again* (Cambridge, MA: MIT Press, 1997), 194–5 and 200–1. See also Clark's chapter on language.
67. Abner Shimony, *Search for a Naturalistic World View*, vol. 1: *Scientific Method and Epistemology* (Cambridge, MA: Cambridge University Press, 1993), 24.
68. Hilary Putnam is quoted in Shimony, *Search for a Naturalistic*, vol. 1, 23: 'What I am saying, then, is that elements of what we call "language" or "mind" penetrate so deeply into what we call "reality" that the very project of representing ourselves as being "mappers" of something "language-independent" is fatally compromised from the start.'
69. Francis Bacon quoted in Shimony, *Search for a Naturalistic*, vol. 1, 23–4.
70. Shimony, *Search for a Naturalistic*, vol. 1, 24 and 31.
71. See, for instance, the following quote from Ernst Cassirer, ‚Der Begriff der symbolischen Form im Aufbau der Geisteswissenschaften' in *Wesen und Wirkung des Symbolbegriffs* (Darmstadt: Wissenschaftliche Buchgesellschaft, 1956), 184–5: 'Selbst bei Kant schien der Schwerpunkt der Lehre mehr in dem, was sie als negative Konsequenz in sich schloß, als in ihrer neuen positiven Grundeinsicht zu ruhen. Als Kern seines Gedankens erschien nicht sowohl der Nachweis, wie die echte Objektivität der Erkenntnis in der freien Spontaneität des Geistes begründet und in ihr gesichert sei, als vielmehr die Lehre von der Unerkennbarkeit des "Dinges an sich".'
72. Cassirer, 'Form and Technology', 24.
73. Ihde mentions Hubert Dreyfus, Patrick Heelan, Robert Ackermann and Ian Hacking. Don Ihde, *Instrumental Realism: The Interface between Philosophy of Science and Philosophy of Technology* (Bloomington: Indiana University Press, 1991), xii.
74. Ihde, *Instrumental Realism*, 100.

4
The Struggle of Titans

Ernst Jünger and Ernst Cassirer: Vitalist and Enlightenment Philosophies of Technology in Weimar Germany

Frederik Stjernfelt

When you read the planned fourth volume of Ernst Cassirer's *Philosophy of Symbolic Forms* (*Philosophie der symbolischen Formen IV*, written around 1928–29, but only published as the first volume of his *Nachlass* in 1995), you find an acute diagnosis of the philosophical situation in Germany and indeed the whole of Europe at the time. The main opposition to Cassirer's own position and the broad neo-Kantian tradition to which he belongs is taken to be the *Lebensphilosophie*, vitalism – here analysed and attacked in its different shapes in Klages, Scheler, Bergson and Heidegger. At the famous Davos confrontation between Cassirer and Heidegger in 1929, this opposition between the tempered rationalism of neo-Kantianism and the decisionist irrationalism of vitalism disguised as phenomenology naturally forms the axis of the discussion: the mature *Bildungsbürger* Cassirer bases his lecture on his recently developed doctrine of 'symbolic forms' – a sweeping generalization of the neo-Kantian emphasis on the philosophy of science to cover also language, myth, the arts and so on – while the young rebel Heidegger places himself on the side of life, with his blending of phenomenology with philosophy of life in his doctrine of the hard destiny of the existence of *Dasein* which is able to realize this destiny only in certain 'Spitzenaugenblicke' in life.

Not long after this meeting, Cassirer expands his doctrine of symbolic forms to cover technology, in his innovative piece about the philosophy of technology presented in the long 1930 essay 'Form und Technik' ('Form and Technology'). Heidegger's mature philosophy of technology, of 'das Gestell' ('the Enframing'), had to wait until after the war – but one of the main influences of that philosophy, Ernst Jünger's book-length essay *Der Arbeiter: Herrschaft und Gestalt* (*The Worker: Dominion and Gestalt*) was also developed at the turn of the decade to appear in 1932.[1] Jünger was indeed an even more staunch vitalist than his friend Heidegger – and the two of them form, together with Carl Schmitt, the leading figures of what was later called 'the conservative revolution', the right-wing philosophy of life of 1920s Germany, to some extent intermixed with or inspiring early Nazism. Cassirer's and Jünger's philosophies of technology thus give us a dual picture of the approach to the pressing issue of technology during the intensive modernization of the interwar period in two of the major currents of early twentieth-century thought: neo-Kantianism and vitalism. Jünger's brand of vitalism is interesting to the extent that he – unlike much vitalism inimical towards modernity and civilization – does not argue against technology, quite the contrary. This may even give rise to hasty identifications between the two Ernsts; thus Andreas Luckner says: 'The central question in Ernst Cassirer or Ernst Jünger on the one hand, in Oswald Spengler and Lewis Mumford on the other, was whether "the" technology was the basis of culture or the enemy of culture.'[2] It is correct that the two of them agree to the degree that none of them participates in the nostalgic and reactionary rejection of technology which was widespread in the period. But, apart from that, their appreciations of technology differ widely, to say the least.

In this chapter, my aim is to compare the two in order to profile Cassirer's analysis of technology. I begin with Jünger's strange and foreboding doctrine.

Technics as Titanic destiny

Ernst Jünger served as a soldier during most of World War I, was wounded many times and received the highest German decoration 'Pour le mérite'. His breakthrough, also to an international audience, was his famous 1920 war diary *In Stahlgewittern*, related in a cool, detached voice. The war experience shaped Jünger's general perception: to him, World War I formed the historical decline and fall of the bourgeois epoch in history, which signalled what would come next, after the decisive 'intrusion of elementary powers into bourgeois space' as he called it.[3] The years around 1930 saw him extrapolate this war doctrine to more general claims, in the ecstatic prose poems of *Das abenteuerliche Herz* (1929/38), in the essays 'Die totale Mobilmachung' and 'Über den Schmerz' (1930 and 1934) and, most thoroughly, in the book *Der Arbeiter: Herrschaft und Gestalt* (1932). The German concept of 'total mobilization' forms the entrance to his doctrine of technology. The idea is that World War I warfare – wherein, of course, the origin of the concept lay – showed that war was no longer a special task for armies fighting it out on battlefields remote from ordinary life; war was rather a test of life and death for the whole of the society and state involved. The country that is able to mobilize, to the most total degree, production, organization, propaganda and dedication will be the one to win the war. This, of course, holds consequences for peacetime also: in order to prepare for war, modern societies must already, during peace, enter a state of total mobilization, involving mass industrialization and coordination on a hitherto unprecedented scale of intensity, range and planning. In *Der Arbeiter*, this idea is interpreted as the outer sign of a major historical destiny deeply affecting modern bourgeois democratic societies, busily trans-forming them into a completely new type of state – the 'Gestalt of the Worker' hinted at by the book title. Jünger's idea is that the worker is no historical product, much less a historical contingency; rather the worker is an overall whole, a gestalt, which is appearing as the overarching structuring principle of a new period about to be born. Everything will now be measured on whether it participates in the worker gestalt or not. This is evident in frontline experience as a drive towards disillusion-ment, de-individualization and objectification: 'In this landscape, where the individual may be discerned only with great effort, fire has burnt out everything which does not have an objective character.'⁴ The difference between working and fighting tends to evaporate within the confines of total mobilization, and the worker enters into the line of command and receives orders on a par with the warrior. Total mobilization is thus a raging, unstoppable process quickly transforming society, destroying bourgeois values, individualism and safety, without yet articulating a new set of values. The current period is thus a chaotic nihilist intermedi-ary, before the worker's gestalt has fully emerged to reveal itself, before order will be restored and a new set of values take over: 'We live in one of those rare epochs where no power rules anymore and no power yet rules.'⁵ This interim period is, at the same time, the period of technolo-gization. To Jünger, the issue of technology is thus tightly connected to his conception of the worker gestalt, and his doctrine of technology is developed within the framework of *Der Arbeiter*. It is characteristic that the single technician does not himself realize his role in this process: he remains closely tied to his specific technical task while the totality of work and the overall technologization process escapes him. In general, human beings do not have any direct relation to technology at all, which is why all attempts to explain technology from such a relation – be they optimist or pessimist, seeing the human being as the creator or the victim of technology – invariably fail.[6] Rather 'Die Technik ist die Art und Weise, in der die Gestalt des Arbeiters die Welt mobilisiert' – Jünger's frequently repeated basic doctrine: technology is the way in which the worker gestalt mobilizes the world. The decisive issue will then be to what degree man is able to represent this worker's gestalt. Technology is the language of the working world, a general language with both grammar and metaphysics, which is why any specific piece of machinery is just as secondary as man: both of them are merely organs for this language to express itself. Here we find an almost phenomenological bracketing: man and machine are bracketed to the benefit of the connection between the two – which is technology as the language of the working world. This technological world thus has its own laws alien to human understanding, just like armament plans may have deeper purposes unknown to the understanding of the planner. History is thus seen as driven by its own deep forces and motives, beyond human control and to a large extent also beyond human understanding. Earlier human positions and connections outside this world – such as bourgeoisie, Christians or nationalists – are attacked by technology which inevitably destroys anything which goes against total mobilization. This process thus forms the decisive historical event of our age, surpassing in importance even the French Revolution and the Reformation.[7] All special privileges fall away during the process; the bourgeois way of life, in particular, loses hope every day because of the introduction of elementary powers (war, revolution, inflation, technology and so on) into the apparent security of bourgeois society, finally to destroy that society completely. To that extent, Jünger's doctrine clearly forms part of what has been called 'the ideas of 1914' as against the 'ideas of 1789' – setting conservative and collectivist values up against the individualist and universal ideals of the Enlightenment. But these conservative values are not simply reactionary demands for a return to pre-1789 principles of religion and politics; rather, they brusquely finish off any remnants of such values as well: technology destroys any religion and forms the strongest anti-Christian power yet experienced. 'Where technical symbols appear, space is emptied of all other forces, of the world of larger and lesser spirits which have established themselves there.'[8] This does not mean, however, that the worker is without faith; rather, he marks the return to a world in which life and cult are again identical ('Leben und Kultus identisch sind').[9] The values to (re-)appear are thus vitalist – which is why greater piety is now observed in movie theatres or at motor races than in churches and temples. Jünger's idea is thus that a new cult is rapidly developing within technology which at the same

time releases and revitalizes ancient forces and principles. In this doctrine, he is at once futurist and conservative and thus a leading member
of the 'conservative revolution', taking the return to earlier values not
to refer to pre-1789 dynasties or clergies, but to much more ancient,
even eternal vital principles. In this process, the faith of the nineteenth
century, that of progress, is also doomed to destruction.[10] The bourgeois
conception of technology – as an organ for progress, leading to virtuous
perfection, closely connected to the values of knowledge, ethics, humanity, comfort – is only one face of technology's Janus head, the other face
being the fact that all technologies possess, as Jünger stated in *Das abenteuerliche Herz*, their own war potential ('potentiel de guerre'). A locomotive may pull a diner wagon as well as military troops, and the chemical
production of nitrates may be used for fertilizing fields as well as for
explosives.[11] This is why total mobilization accelerates ever stronger
during technologization, even if it appears superficially peaceful; it
inevitably pulls both individuals and societies into the worker-soldier's
way of life. The war thus displays the power of technology in its pure
form, beyond any connection to progress or to economy. Hence, technology is no neutral stock of practical means for different purposes;
rather, it fundamentally changes the whole lifestyle of those who adopt
it. The field which is fed by artificial fertilizer is no longer the same field,
and the peasant cultivating it has become part of the worker's gestalt[12] –
an example which Heidegger was famously to elaborate on[13] – and the
difference between urban and rural areas evaporates along with the
peasant lifestyle: 'The man who admits an electrical connection may
achieve greater comfort but also less independence than he who lights
his oil lamp.'[14] Jünger's doctrine thus also forms one of the early accounts
of globalization: the nation that makes use of machines, engineers and
special workers enters into a visible or invisible tribute relation which
blows all traditional connections apart. The result – not unlike Marx's
analysis of capitalism – is anarchy. But the anarchical result only forms
a specific phase in Jünger's account. This is because anarchy only forms
the prerequisite for the formation of new structure: 'This anarchy is
nothing but the first necessary step leading to new hierarchies.'[15] The
atomizing of former structures actually diminishes the resistance against
becoming part of a new 'organic world construction'.[16] It is almost a
positively turned version of Hannah Arendt's famous hypothesis that
modern totalitarianism is possible only on the basis of the atomization
of individuals in bourgeois society; in Jünger, the new totalitarianisms
are accepted, if not greeted, as facts and 'In technology we acknowledge
the most efficient and most irrefutable means of total revolution.'[17] It is

important, moreover, to note the emphasis on new 'organic' constructions: Jünger's conception of technology is not mechanistic; rather, the deep involvement of the worker in the process marks the transgression of any distinction between mechanic and organic structures. Understanding is unable to grasp this development because it remains tied to space-time representation, Jünger maintains – the worker's gestalt being a structuring force beyond space-time. This tie to space-time representation and the idea of progress is, at the same time, the reason why bourgeois understanding naively imagines technology will develop indefinitely. It will not.[18] The process of technologization will only continue until it has provided tools for the 'special demands which the worker's gestalt subjects it to'[19] – demands which we still do not know because the process has not yet ended to reveal its purpose. This is why contemporary land- and cityscape look like workshops and our era is unable to construct lasting monuments. Urban environments are split between museums, protecting and displaying remnants of former, orderly times, on the one hand, and constantly changing industrial areas on the other. At the same time, technology approaches perfection, and the day it is achieved, technology development will come to a complete halt. Then there will be no technical service which is not connected to other such services, and the sum of special working properties will constitute one total working gestalt; all of society will be welded together into one gigantic instrument.[20] The whole process takes the same overall character as the growth of a plant, where only the mature organism allows us to grasp the reason for the earlier phases of its development. Another metaphor of Jünger's to express this development is that of a pyramid, where the four sides are under construction and where the finishing summit can be expected within a foreseeable future.

Jünger's understanding of man and machine as mere side effects of the technologization process, of course, leaves the issue of what effect this process will have on life. Nobody would learn any definite skill, everybody would remain in education, everybody would have to run faster and faster. It is already the case, just like in battle, that these demands result in many victims – but they must be seen as martyrs for the heightening and intensifying of life to a new level of organization, where the present, chaotic and revolutionary space will give way to a new, highly ordered space. This end of technology development thus coincides with the appearance of a new type of total political power where ordered life, according to the gestalt, becomes possible, and the planned economy will form part of the structure, as this new order would follow different economical and business laws from those currently in place. Technology will be perfect and easy to understand (it would automatically include plant and animal parts); long-lasting calculability will erase the experimental qualities of present technology and it will again become possible to erect monuments.[21] This new order demands no less than global validity.[22] Technology is thus no separated causal system; rather, it is intrinsically connected to social power.[23] As against the more traditionally conservative scepticism regarding technology, Jünger's doctrine identifies technology and nature: 'Technics and nature are not opposites – if they are experienced like that, it is a sign that life is not in order.'[24] As technology possesses its own laws, it is beyond human power to make decisions to influence the end of technology; all we can do is to sacrifice ourselves by throwing ourselves into the process in order to speed it up as much as possible: 'it is rather our task to increase the force and velocity of the processes, in which we are seized'.[25] By such sacrifice, the new society at the end of the process will come nearer.

Jünger leaves nobody in doubt that this process is also one with immediate political consequences: 'The entry into imperial space presupposes a testing and a hardening of the landscapes of planning, of which we may not even today have any idea,'[26] leading to new kinds of states beyond any present comparison. 'It is, however, possible to predict that neither work nor democracy, in the ordinary sense of these words, will then be spoken about.'[27] Jünger is unequivocal in his description of the totalitarian, worldwide workers' empire about to be realized. Even if the description borrows much from Soviet Bolshevism and Italian fascism – and from German Nazism still reaching for power – direct references to these political projects are absent in the book's more general, divinatory descriptions. The book's project is, rather, to distill the essentials of these developments and make a prognostic of their general direction. In the destiny-ridden language of the book, this direction is depicted more like an uninterruptable process than a political choice *pro et con*, which is also why its last lines make the following dark demand: 'It is impossible, without being touched, to observe how human beings are occupied, in the midst of chaotic zones, by the readiness to use weapons and the hardening of hearts, and how they can reject the pathway of happiness. Here, the effort which is expected of us is one of participation and duty.'[28]

Technology as symbolic form

Cassirer's account of technology makes up part of his mature philosophy of symbolic forms. At the same time, it occupies a strange, stepdaughter-like place in that doctrine. While myth, language and science appear as prototypical symbolic forms and thus occupy one volume each of Cassirer's *Philosophy of Symbolic Forms*, other symbolic forms like art, politics and religion are treated less thoroughly. Technology counts among them, being only dealt with in the 1930 essay 'Form and Technology'. And even this essay deals, to a large extent, with the refutation of received conceptions of technology of the period; it does not include the analysis of any single example of technology. To that extent it falls prey to phenomenology's criticism of neo-Kantianism in general: examples are strangely lacking, as Husserl once wrote about Heinrich Rickert: 'one does not find one single example – and one also does not miss one'.[29] The latter is not quite true, though; in a field as wide as technology, ranging from quill pens to telegraphs, from flint axes to movie theatres, analytical examples would surely strengthen the account. Cassirer's paper remains, though, a rare and early example of an attempt at positively defining technology. Let us run through its main ideas. For a start, Cassirer is in agreement with Jünger about the spread and centrality of technology in present culture and society, measured in terms of its actual workings and services: even its enemies are now forced to make use of it. Vitalism, itself, becomes subject to the forces of technology – even in the cases when it most decidedly opposes it.[30] But if technologization may not be contained or stopped, like the case in Jünger, to Cassirer it does not call for subservience and duty – quite the contrary. His main contention is that technology crucially and in a new way poses the question of the freedom of the human mind. But philosophy has not yet investigated the meaning and legitimacy of technology, its origin and validity, as it has other domains of culture – due to its seemingly peripheral character. The central task is thus to determine its place in relation to other such domains. In line with Cassirer's overall philosophy, the sequence now followed is, first, to investigate its meaning – the idea which motivates it – and from there proceed to the issues of its being and legitimacy.[31] Cassirer here refers to the remarkable fact that, in devising his doctrine of ideas, Plato built on technological examples rather than natural ones. The idea of a technological device is not an imitation of an existing entity; rather it is only accessible via grasping its *eidos* and *telos* in order to bring it into existence.[32] Thus, the technological invention is aimed at a general form which is what provides the essence of that invention. Just as he emphasizes the necessary materiality of language, he emphasizes the ideality of technology: both are material systems invested with ideality – to put it one way, they are 'extended minds' – an ideality defined by the invariance of different material instantiations of the same sign or machine.[33]

Cassirer finds accounts of this ideality of technology in contemporary philosophers like Friedrich Dessauer and Max Eyth – and this is why it must be distinguished from natural science with which it is very easily confused, as when speaking about 'applied science'. In Jünger, the identification between the two was the other way around: science is not-yet-constructed technology. To Cassirer, the important issue is to find the relation of technology to the totality and universality of human spirit, and this is only possible 'when we focus on the *concept of form* rather than the *concept of being* as understood in the natural sciences',[34] the form of technical working. But form and meaning must here be distinguished sharply from value. Cassirer here argues against a tradition, originating with Rousseau, of (vulgar) pragmatism, focusing upon pleasure and utility and forcing technology to be judged by criteria foreign to itself. Technology must be judged in its functioning, in becoming; it must be given a genetic definition.[35] Cassirer thus opts for a functional understanding of technology, related – but not identical – to his famous functional theory of science developed from *Substanzbegriff und Funktionsbegriff* (1910) onwards. The idea here, of course, was that the object of science is not the substance of things, but the invariances in functional interrelations between entities.[36]

Max Eyth's claim that language and technology were man's two original faculties (Eyth tends to praise the former and forget the latter) points for Cassirer to a genuine problem, because language always had a deep-seated instrumental meaning along with its descriptive and theoretical meaning: the basic duplicity in understanding of thought and action.[37] Here, Cassirer sharply distinguishes between magic and technological mastery of nature, and consequently argues against Frazer's idea of magic as primitive technology. In magic, man is not a scientist with limited tools; rather man falls victim to his own immediate desires – he has not yet made the stable distinction between subject and object which is developed in phase of *Darstellung*, representation, in specific ways in each of the symbolic forms. In magic, man has substituted his subjection to his own immediate desire to shape the world for his subjection to immediate expression. By contrast, Cassirer's claim is that technology requires quite the opposite: the distancing of immediate desire. Desire is preserved – but kept at a distance which allows the finding and construction of a governed outlet for its impulses.[38] Only when the goal is placed in a remote position does an objective experience become possible[39] in which nature is represented as an independent realm of being with its own structures and powers which you must be subject to if you want to use them: 'Human beings no longer attempt to make reality amenable to their desires with various methods of magic and enchantment. They take it as an independent and characteristic "structure".'[40] Reality is represented in such a way that its plasticity is placed in a frame determining certain rules for the possible outcomes. The first tool is, at the same time, the first mediation, making possible also understanding, characterized by indirect mediation between beginning and end: the tool constitutes, in itself, a 'terminus medius', an intermediate concept.[41] Thus learning intermittently to look away from his goal, man achieves the ability to reach it, unlike the animal remaining bound by instinct to certain procedures. Here, Cassirer invokes Kant against Frazer: causal understanding requires the concept of necessary connection, and not merely Frazerian associations sufficient for constituting the realm of magic. The tool presents a tie between different objective determinations, that of the tool and that of its object, grasped in the synthesis of functional unity. These ideas, possibly influenced by Heidegger's analysis of 'Zuhandenheit' ('readiness-to-hand') in the first part of *Sein und Zeit* (*Being and Time*), result in the claim that, in the tool world, there are no mere things – rather a whole of mathematical vectors. Cassirer thus situates technology in an organic holism not unlike Jünger. But while the latter's holism tended to embrace the whole of technological society, thus precluding the initiative of man, Cassirer's organic holism remains on a much smaller scale, compatible with human freedom. Both language and tools easily appear, now, to primitive man as a foreign pandemonium he has not himself created – but in Cassirer's account of the Enlightenment process man gradually realizes more and more that he is, in fact, its free creator; a new subjectivity of Titanic pride and freedom emerges as the correlative of technological development.[42] Quite unlike Jünger, who saw the Titanic as referring to the worker's gestalt, which increasingly subjects the individual to a new totalitarian order, Cassirer sees the Titanic in man himself, gradually acquiring greater mastery over technological development. Thus, technology is what makes possible the distinction of human *will* from mere desire and expression of wishes, as emphasized by John Michael Krois.[43] Technology, from this perspective, thus forms an overlooked prerequisite to central human phenomena like ethics and individual personality. The will is, unlike mere wishes, subject to ethical judgment – because the distance between the will and its object, means and ends, makes possible the distinction between different wilful series of actions. Similarly, defining personality as the consistency of action,[44]

Cassirer makes the symbolic form of technology central to the development of individuality. The functional joining together of objects with respect to a goal is central, both for the external construction of technology and the individual's construction of its habitual action and, thus, of the individual itself. Here, Jünger's idea that technologization necessitates the de-individualization, standardization and hardening of persons to the extent that it calls for the creation of a new type of human being[45] points in the exact opposite direction. In Cassirer, the gradual development of technology not only gradually extends human freedom but participates in the unfolding of individuality, its critical and ethical self-control.

Cassirer versus vitalist criticism of technology

Of course, Cassirer immediately needed to address the vitalist worries, widespread at the time, regarding the possible alienation of man in the face of the increasing technological productions. Jünger's interpretation of technology had not yet appeared, so Cassirer could not address that. It is doubtful, however, that he would have done, even if he had known about it (see below).

But towards the end of the paper, Cassirer, over and over again, confronts the pessimist claims in criticisms like those of Ludwig Klages, Georg Simmel, or Walther Rathenau, in order to make his own final position clear. It is interesting to note here that, apart from Klages (who most decidedly belonged to the right wing), Cassirer mainly concentrates his discussion on versions of vitalism which do not belong to the right wing – those of the two Jewish democrats, Simmel and Rathenau (the latter a liberal and the former probably a social democrat). This permitted him to appreciate aspects of their criticism and ascribe to it an – albeit restricted – validity before proceeding to his counterattack. As the articulation of Cassirer's final position is filtered through his discussion with these vitalists, let us take them one by one.

The vitalist Ludwig Klages, the detractor of 'logocentrism', is here typecast as a cultural pessimist, protecting the vegetative soul against the destructivity of technological mind and spirit. Cassirer may seem surprisingly easy in admitting the obvious destructive effects of technological spirit. As a cure, he here points to Schiller's concept of *Spieltrieb*: only in playing and creating, man himself is created as such.[46] But to Cassirer, all of the symbolic forms take part in such creative processes: 'Each new gestalt of the world opened up by these energies is likewise always a new opening out of inner existence; it does not obscure this existence, but makes it visible from a new perspective.'[47] In a double movement back and forth, every symbolic form makes a new kind of subjectivity crystallize on the part of man. Here, Cassirer refers to Ernst Kapp's interesting idea that only through technology man becomes able to understand the workings of his own organs.[48] Technology provides a special dimension of man's understanding of himself; even if technology begins as the simple prolongation of man's bodily abilities, it quickly loses these ties and may develop solutions working wholly differently from nature's bodily functions: the aeroplane uses different flying techniques than the bird flapping its wings. Cassirer is by no means unreceptive, however, to the complaints of vitalists that technology does in fact drive out man from a paradise of immediacy: 'The moment the human being devotes himself to the hard law of technological work, the abundance of immediate and unbiased happiness that organic existence and activity had given him fades away forever'[49] – as if that life was not also solitary, poor, nasty, brutish and short, as Hobbes would have it.

More precisely, the problem inherent in technology is that tools obey their own laws, which break into the free rhythms of organic life[50] – Cassirer's version of his old teacher Georg Simmel's 'tragedy of culture' that man invents more and more things which form external constraints turning against the world of the subject. Technology follows its own laws, alien to simple, organic existence. And, what is more, the tragedy also stems from the danger that technology may overstate its aims and infringe on those of other symbolic forms. Thus, its relation to them must also be settled. This is not, of course, a task which may be completed within the scope of this single article, but Cassirer makes some initial inroads into this problem, especially in distinguishing technology from the symbolic forms of science and art, with which it shares some properties; this should not prompt the identification of them nor relations of dominance between them. Science and technology as symbolic forms constrain and support each other – Cassirer quotes Goethe's pragmatist words on the unity of thought and action; their decisive difference lies in the former's reference to what is, and the latter's to what can be done, in cases where that answer may only be given in the shape of the invention and construction of certain artefacts.[51] The technician, as it were, is a demiurge who selects among existing possibilities; he must find the solution in the world of ideas and transport it into this world.[52] Cassirer thus escapes the fateful tendency in much twentieth-century German thought indiscriminately to identify science and technology.[53] On the other side, Cassirer distances himself from a too close an identification of technology and art, even if creativity and aesthetics do take part in both of them, when form celebrates its victory over matter. Art, however, remains tied to subjective expression and the synthesis between the objectivity of the artwork and the subjectivity expressed therein, whereas creativity plays another, more objective role in technology. The bottom line remains that even if technology does indeed have rules and principles of its own, its overall place in the interplay of symbolic forms must be decided by which general purposes it serves in human development.

In the overall perspective, Cassirer, of course, takes the side of technological mediation, the main arguments against the Simmelian criticism being two. First, Simmel's alleged tragedy of culture remains indiscriminately the same in all areas – language, science, arts and the like; all of them are, just like technology, objective devices created by us and subsequently also constraining us, and it would obviously be meaningless to imagine a serious political resistance against all such forms. Technology only differs from the other symbolic forms in displaying this fact more evidently. Second, and most decisively, the basic measuring stick of all symbolic forms must come not from life, not from pleasure and displeasure – but from freedom and lack of freedom.[54] Thus, Cassirer's final vindication of technology rests on the fact that it does, in fact, increase human freedom; if it led to still more subjection or slavery, it should be discarded. In Jünger's conception, by contrast, that is exactly the predicament of modern technology. This forms Cassirer's decisive argument: it is indeed correct, as the vitalists claim, that technology forces man away from any idea of a life immediately and organically embedded in nature. But the vitalist criteria, are not sufficient. The more basic critieria to be used are those of the human mind (*Geist*), implying the renouncement of simple organic pleasure in favour of freedom. This is how man is cast as a Titan, not without heroism: 'For the path of mind stands here as everywhere under the law of renunciation, under the command of a heroic will that knows it can only reach its goal through such renunciation, establishing itself through it and renouncing all naive and impulsive longings for happiness.'[55] Freedom is not easily reached, and Cassirer here points to the fact that the achievement of freedom must renounce other criteria like that of simple pleasure or happiness, however tempting they may appear.

And this is why Cassirer's complex conclusion may seem to vacillate between a pessimist criticism of technology and the claim that technology, in return, brings with it new aspects of subjectivity. In a final confrontation with the vitalist criticism, now in the shape of Walther

Rathenau's attacks on the technical culture,[56] Cassirer initially goes far in embracing his sombre views: 'Whoever does not from the beginning subscribe to the demands of simple utility and instead treasures the meaning of ethical and spiritual standards, cannot carelessly pass over the grave inner damages of a lauded "technological culture".'[57] He paraphrases Rathenau as follows: 'On one hand, there is completely soulless and mechanized work, the hardest chore. On the other hand, there is unrestricted will to power and will to rule, unrestrained ambition and meaningless consumerism.'[58] Might not all this be read just as Cassirer paraphrasing Rathenau, only in order to refute him – or does he, to some degree, agree with Rathenau's verdict? When introducing Rathenau, Cassirer unambiguously embraces this criticism: 'Few modern thinkers have as keenly observed and forcefully uncovered this damage as Walther Rathenau. He has done so with growing zeal and passion in his writing.'[59] This technological desert landscape creates an empty hunger of pure hedonism: 'Every satisfied need serves only to bring forth new needs in increasing measure – and, once you have entered it, there is no escape from this cycle.'[60] After thus admitting the force of traditional conservative criticism of the drawbacks of technology, Cassirer brings forth his central argument: these drawbacks do not stem from the essence of technology, its gestalting principle, but remain mere external effects of capitalism.[61] This idea seems, in fact, to stem from Rathenau,[62] himself an industrialist who, despite of his pessimist attacks on technology, maintained the inevitability of the process of technologization. Rathenau found its dark implications could be mended by rationalization, of which he was one of the early great proponents, both in theory and practice, on the one hand, and state intervention against capitalist excesses, on the other. Cassirer's final distinction between the spirit of technology and the spirit of capitalism thus explicitly stems from Rathenau. Cassirer now develops further the idea from the confrontation with Klages that the development of technology has its goal not only in subjecting the forces of nature to the service of mankind, but also in subjecting the chaotic forces of man himself to a similar rule, thereby raising and shaping the will and the ethics of work. In this interpretation, technology thus also contains the cure against capitalism. The idea is that technology as such creates a community of destiny urging solidarity between the people participating in it[63] and thus that its outward thrust is supplemented by its evolution of a new form of subjectivity, in relation to whose purposes technology must remain a servant.

Edward Skidelsky (2008) rightly observes that this hopeful conclusion is indeed vague – and goes on to claim it proves the overall failure of

Cassirer's attempt at including technology in his philosophy. Cassirer's support for Rathenau's acerbic anti-technological observations are not convincingly argued away by the reference to solidarity; rather they prove that Cassirer's principle that each symbolic form should be judged by its own internal criteria only, or at least primarily, founders in the meeting with technology. For his refusal to accept Rathenau stems from an ordinary moralist, anti-hedonist critique, not in any way from criteria specific to the symbolic form of technology, according to Skidelsky.[64] But Cassirer's quasi-Marxist pointing to the emergence of a technological solidarity, however vague, does not fall prey to Skidelsky's criticism. For Cassirer's analysis of technology as the joining of objects into a functional whole might have its subjective correlate not only in the individual's articulation of and control of his will (see Krois above), but also, intersubjectively, in the articulation of a functional whole of interacting, creative human beings. Such a joining together in a functional whole, however, might easily acquire the scent of totalitarianism. Here, Jünger's parallel pointing to technologization as the joining together of human beings into an overarching functional whole permits us to see the difference. The whole in Jünger has its own, deep laws, alien to human understanding, and its totality on a global level precludes any individual influence upon it – indeed, serves to eradicate what may have been left of such influence. Cassirer's functional whole differs from such views in two ways: even if it also has its own rules, alien to immediate organic pleasure, it is no global destiny, but tied to the range and effect of each single piece of technology. Even if Cassirer does not himself undertake analyses of specific pieces of technology in 'Form and Technology', his conceptual apparatus opens the possibility of such analyses in terms of the specific *Gefüge*, the specific part–whole articulation of the functioning of the human–machine interface in each single case. Thus, unlike Jünger and the vitalists, Cassirer's conception does not preclude human freedom, but rather emphasizes the possibility of specific further developments of that freedom, both in terms of new action types being made possible by technology and, more broadly, in terms of participating in the further, collective development of technology.[65] Seen from a Cassirerian point of view, Jünger's doctrine, by contrast, is an example of mythical consciousness, an ideological relapse subjecting man to dark forces beyond any possible control: from the point of view of Cassirer's later, political work, *The Myth of the State*, it forms part of the fateful return of mythical thought in mid-twentieth-century politics. Its claim for the power to predict the evolution of human history, moreover, contrasts with Cassirer's refusal to admit the possibility of such prediction based in human liberty.

Cassirer was politically a social-liberal and, during the Weimar years, adherent of the small centre party Deutsche Demokratische Partei. By contrast, Jünger was an aristocratic supporter of the far right wing and, even if he consistently avoided Nazism, his theory of state undoubtedly forms a variant of fascism. The duplicity of social liberalism is nowhere as apparent in Cassirer's philosophy as in his final appraisal of technology. He admits technology does, in fact, bring with it new constraints and alienations, as indicated by the vitalists – but the bottom line remains positive because these drawbacks are the flip side of what makes the further development of man and human freedom possible. Thus, in the larger perspective, technology remains the servant, human freedom its master.

The Enlightenment character of this conclusion builds on Cassirer's choice of human freedom as the final measuring stick of symbolic forms. Hedonist qualities like pleasure and displeasure may, of course, play an important role in the immediate appreciation of a piece of technology, and may participate in the evaluation of it – but they must not serve as the final arbiter of technology as such; human freedom remains, in all cases, the most noble criterion because of serving as the condition of possibility of the further development of symbolic forms. Other criteria may stop or want to obstruct that development (like the criteria of order and new values in Jünger) and are thus reconcilable with societies where, for example, pleasure is taken as a basic criterion at the price of human freedom. Cassirer's argument seem to rest on the idea that liberty is to be conceived of as a metaprinciple, allowing for the free play of other value principles such as pleasure, happiness, order, honour, recognition, wealth, health, beauty and so on, developed and supported by different symbolic forms. Only liberty allows for the ongoing free articulation of other criteria with and against each other.

This explains the basic difference between a vitalist and an enlightenment philosophy of technology, despite much agreement. As to Cassirer and Jünger, they agree in the attempt to find a positive definition of technology; they agree on technology having rules and principles of its own, alien to man's immediate understanding; they agree on many of the unwanted implications of technology; they even agree on the necessity of accepting these implications. Still, their accounts, even if the recognizable products of one and the same Weimar culture, are worlds apart. While the futurist vitalism of Jünger finds the Titan in the overarching technological process in which man remains one gear wheel among many, the Enlightenment stance of Cassirer locates the Titanic in the ongoing development of man and human freedom, struggling against demands to bow to alien deities.

Notes

1. Heidegger's voluminous notes and sketches pertaining to Jünger (especially *Der Arbeiter*) are collected in Martin Heidegger, *Gesamtausgabe*, vol. 90: *Zu Ernst Jünger*, ed. Peter Trawny (Frankfurt am Main: Klostermann, 2004). Heidegger conceives of Jünger as an acute observer who takes Nietzsche's will-to-power metaphysics as the basis for his detailed description of modern society. Thus, the figure of the worker is Jünger's version of the Nietzschean Übermensch. Jünger, however, lacks a philosophical interpretation of his observations.
2. 'Die zentrale Frage etwa bei Ernst Cassirer oder Ernst Jünger auf der einen, Oswald Spengler und Lewis Mumford auf der anderen Seite war, ob 'die' Technik Grund oder Feind der Kultur sei ...' (English transation my own). Andreas Luckner, *Heidegger und das Denken der Technik* (Berlin: Transcript, 2008), 32.
3. Ernst Jünger, *Der Arbeiter: Herrschaft und Gestalt* (Stuttgart: Klett-Cotta, 1982 [1932]), 48.
4. Jünger, *Der Arbeiter*, 111. 'In dieser Landschaft, in der der Einzelne nur sehr schwer zu entdecken ist, hat das Feuer alles ausgeglüht, was nicht gegenständlichen Charakter besitzt.' The English translations of this and the following quotes from *Der Arbeiter* are my own.
5. *Ibid.*, 190. 'Wir leben in einem der seltsamen Zeiträume, in dem Herrschaft nicht mehr und Herrschaft noch nicht besteht.'
6. *Ibid.*, 155–6.
7. *Ibid.*, 157–8.
8. *Ibid.*, 161. 'Wo die technischen Symbole auftauchen, wird der Raum von allen andersartigen Kräften, von der großen und kleinen Geisterwelt, die sich in ihm niedergelassen hat, entleert.'
9. *Ibid.*, 162.
10. As against the supposedly linear notion of progress, Jünger thus posits a cyclic conception of time where processes of destruction and construction eternally interchange.
11. Jünger, *Der Arbeiter*, 163.
12. *Ibid.*, 167.
13. In his comparison of the production of corpses in concentration camps with industrialized agriculture, see Martin Heidegger, 'Einblick in Das Was Ist', *Gesamtausgabe*, vol. 79: *Bremer und Freiburger Vorträge*, ed. Petra Jaeger (Frankfurt am Main: Klostermann, 1994), 27: 'Ackerbau ist jetzt motorisierte Ernährungsindustrie, im Wesen das Selbe wie die Fabrikation von Leichen in Gaskammern und Vernichtungslagern' ('Agriculture is now a motorized nutrition industry, in essence the same thing as the production of corpses in gas chambers and extermination camps'; my translation).
14. Jünger, *Der Arbeiter*, 167–8. 'Der Mann, der sich einen elektrischen Anschluß legen läßt, verfügt vielleicht über größere Bequemlichkit, sicher aber über geringerer Unabhängigkeit als der, der seine Lampe brennt.'

15. *Ibid.*, 169. 'Diese Anarchie ist nichts anderes als die erste, notwendige Stufe, die zu neuen Rangordnungen führt.'
16. *Ibid.*, 169.
17. *Ibid.*, 169. 'In der Technik erkennen wir das wirksamste, das unbestreitbarste Mittel der totalen Revolution.'
18. Jünger's argument here is curiously akin to the argument in Soviet mathematics against the 'bourgeois' mathematics of set theory and infinity: in the workers' state, such idealist conceptions are supposed to vanish.
19. Jünger, *Der Arbeiter*, 172.
20. *Ibid.*, 175.
21. *Ibid.*, 190.
22. The book does not make a clear prognosis about when this new order should be about to appear, but indirectly you get the impression it is not very far from the moment of writing and that Bolshevism and fascism form part of the overall development described. At the time, it must have seemed tempting to read the January 1933 *Machtergreifung* into the divinations of the 1932 book. In 1995, I had the rare opportunity of interviewing the then 99-year-old Jünger. His overall analysis had not changed: now the notion of worker's gestalt had been exchanged for that of the 'Titans', but the estimated time for the end of the Titanic process of technologization had been extended considerably. By then, he felt that the new order would probably have to await the dawn of the twenty-second century, which is why he claimed: 'I am not much interested in the twenty-first century. It will still be the century of the Titans. But in the twenty-second century, the gods will return.' Nils Gunder Hansen and Frederik Stjernfelt, 'Interview with Ernst Jünger', *Kritik* 114 (1995).
23. Jünger, *Der Arbeiter*, 202.
24. *Ibid.*, 203. 'Technik und Natur sind keine Gegensätze – werden sie so empfunden, so ist dies ein Zeichen dafür daß das Leben nicht in Ordnung ist.'
25. *Ibid.*, 203. '... es gilt vielmehr, die Wucht und die Geschwindigkeit der Prozesse zu steigern, in denen wir begriffen sind.'
26. *Ibid.*, 306. 'Dem Eintritt in den imperialen Raum geht eine Erprobung und Härtung der Planlandschaften voraus, von der man sich heute noch kein Vorstellung machen kann.'
27. *Ibid.*, 307. 'Es läßt sich jedoch voraussehen, daß hier weder von Arbeit noch von Demokratie in dem uns geläufigen Sinne mehr die Rede sin wird.'
28. *Ibid.*, 307. 'Nicht anders als mit Ergriffenheit kann man den Menschen betrachten, wie er inmitten chaotischer Zonen an der Stählung der Waffen und Herzen beschäftigt ist und wie er auf den Ausweg des Glückes zu versichten weiß. Hier Anteil und Dienst zu nehmen: das ist die Aufgabe, die von uns erwartet wird.'
29. Edmund Husserl, *Husserliana*, vol. XXII: *Aufsätze und Rezensionen, 1890–1910*, ed. Bernhard Rang (The Hague: Martinus Nijhoff, 1979), 147.
30. See Ernst Cassirer's 'Form and Technology' in this volume, 16. See also 'Form und Technik', in Ernst Cassirer, *Gesammelte Werke* [hereafter ECW], vol. 17: *Aufsätze und kleine Schriften* (1927–1931), ed. Tobias Berben (Hamburg: Felix Meiner, 2004), 139–83, 140–1.
31. Cassirer, 'Form and Technology', 18/ECW 17, 145–6.
32. *Ibid.*, 19/ECW 17, 144.

33. This is emphasized by Gideon Freudenthal, 'The Missing Core of Cassirer's Philosophy: Homo Faber in Thin Air', in Cyrus Hamlin and John Michael Krois (eds), *Symbolic Forms and Cultural Studies* (New Haven: Yale University Press), 212.

34. Cassirer, 'Form and Technology, 19–20/'statt des *Seinsbegriffs* der Naturwissenschaft, vielmehr den *Formbegriff* in den Mittelpunkt stellt und sich auf *seinen* Grund und Ursprung, seinen Gehalt und Sinn zurückbesinnt', ECW 17, 145.

35. *Ibid.*, 22/ECW 17, 148.

36. See Ihmig's thorough reconstruction (1997, 2001) of Cassirer's philosophy of science with its emphasis on his inspiration from Felix Klein's invariance definitions of different geometries.

37. *Ibid.*, 22–3/ECW 17, 150.

38. It must be said that in observations like these, Cassirer assumes the engineer's point of view, that of the inventor and constructor of technology, rather than that of the ordinary user whose experience will often be a mixture of relief and alienation.

39. This technological objectification thus makes possible the appearance of neutral objects which are not immediately seen as hindrances or supports for desire's goal. In Cassirer's appropriation of the biologist von Uexküll's theory (see Stjernfelt 2009), this issue of the appearance of objects during evolution is important – even if Cassirer vacillates, along with von Uexküll, as to the acceptance of neutral objects also in higher animals.

40. *Ibid.*, 29/'Der Mensch sucht nicht länger, sich die Wirklichkeit mit allen Mitteln des Zaubers und der Bezauberung gefügig zu machen; sondern er nimmt sie als ein selbständiges charakteristisches "Gefüge".' ECW 17, 157.

41. *Ibid.*, 30/ECW 17, 158.

42. *Ibid.*, 34/ECW 17, 163.

43. John Michael Krois, *Cassirer: Symbolic Forms and History* (New Haven: Yale University Press, 1987), 103.

44. *Ibid.*, 104.

45. In 'Über den Schmerz', 1934.

46. Also the mature Peirce appropriated this concept from Schiller, see his 'play of musement'.

47. Cassirer, 'Form and Technology', 37/'Jede neue Gestalt der Welt, die durch diese Energien erschlossen wird, ist zugleich immer ein neuer Aufschluß über das innere Sein – sie verdunkelt dieses Sein nicht, sondern macht es von einer neuen Seite her sichtbar.' ECW 17, 167.

48. Here, one can think of William Harvey's discovery of the function of the heart, characterizing it as a blood pump.

49. Cassirer, 'Form and Technology', 39/'In dem Augenblick, in dem sich der Mensch dem harten Gesetz der technischen Arbeit verschrieben hat, sinkt eine Fülle des unmittelbaren und unbefangenen Glücks, mit dem ihn das organische Dasein und die rein organische Tätigkeit beschenkte, für immer dahin.' ECW 17, 170.

50. *Ibid.*, 39–40.

51. Cf. H. H. Pattee's (1969) insistance that neither organisms nor machinery can be understood in mechanical terms only. Mechanics can account for the physical process taking place in a machine which does not function, just as

well as that of a machine which functions. That is, the concept of functioning is not mechanical.
52. Cassirer, 'Form and Technology', 44/ECW 17, 176.
53. The tendency to identify science and technology is obvious in Jünger as well
as in Heidegger's more traditionally conservative version. But this identification is found not only on the right wing – a left-wing version would be
Habermas's easy equality sign between the two under the headline of instrumental rationality ('Zweckrationalität').
54. Cassirer 'Form and Technology', 41/ECW 17, 172–3.
55. *Ibid.*, 41/'Denn der Weg des Geistes steht hier wie überall unter dem Gesetz
der Entsagung: unter dem Gebot eines heroischen Willens, der weiß, daß er
sein Ziel dadurch zu erreichen, ja daß er es nur dadurch aufzustellen vermag,
daß er auf alles naiv-triebhaften Glücksverlangen verzichtet.' ECW 17, 173.
56. Walther Rathenau, an impressive figure, was a Jewish-German industrialist, philosopher, writer and politician, assassinated when he served as the
Foreign Minister of Germany in 1922 by two ultra-nationalist officers. It is
probably no coincidence that Cassirer chooses to end the paper with exactly
Rathenau to represent the vitalist criticism of technology. Rathenau was a
nationalist, but at the same time a liberal – one of the co-founders of the
Deutsche Demokratische Partei, which Cassirer himself supported – and he
was despised as a Jew by the German right wing, which was behind his death.
His 1922 assasination was, already in 1930, widely seen as a fateful event in
Germany's ongoing development away from Enlightenment norms. By
picking his version of vitalism, Cassirer achieves two things: he separates the
vitalist critique from its primary right-wing position (of Klages, Heidegger,
Jünger and many others), in line with his own partial appraisal and understanding of that position. And, what is more, by doing so he addresses an
important split within the liberal tradition itself, between happiness and
liberty as competing guiding principles – see, for example, Millsian utilita-
rianism versus the Kantian emphasis on freedom. Jünger, on the other hand,
attacks Rathenau in his 1930 paper 'Die totale Mobilmachung'; he gives
him credit for supporting total mobilisation during World War I and urging
for protests against the Versailles treaty, but, after the war, he remarked it
would have been tragic to see the German troops marching triumphantly
through the Brandenburg Gate. To Jünger, this proved he had not really been
penetrated by total mobilization.
57. Cassirer, 'Form and Technology', 48/'Wer sich nicht von vornherein den
Forderungen der bloßen Nutzbarkeit verschrieb, sondern sich den Sinn für
ethische und für geistige Maßstäbe bewahrte, der konnte an den schweren
inneren Schäden der gepriesenen "technischen Kultur" nicht achtlos vor-
beigehen.' ECW 17, 180–1.
58. *Ibid.*, 48/'Völlige Entseelung und Mechanisierung der Arbeit, härtester
Frondienst auf der einen Seite – unbeschränkter Macht- und Herrschaftswille,
zügelloser Ehrgeiz und sinnloser Warenhunger auf der andern Seite ...' ECW
17, 181.
59. *Ibid.*, 48/'Unter den modernen Denkern haben wenige diese Schäden so
scharf gesehen und so schonungslos aufgedeckt, als es Walther Rathenau
mit immer wachsender Energie und Leidenschaftlichkeit in seinen Schriften
getan hat.' ECW 17, 181.

60. *Ibid.*, 48/'Jedes gestillte Bedürfnis dient nur dazu, in gesteigertem Maße neue Bedürfnisse hervorzutreiben – und aus diesem Kreislauf ist für den, der einmal in ihn eingegangen ist, kein Entrinnen.' ECW 17, 181.
61. *Ibid.*, 48–9/ECW 17, 182.
62. *Ibid.*, 49n/ECW 17, 182 note.
63. *Ibid.*, 49/ECW 17, 183.
64. Edward Skidelsky, *Ernst Cassirer: The Last Philosopher of Culture* (Princeton, NJ: Princeton University Press, 2008), 186–90.
65. Cassirer's Enlightenment stance is here remarkably clear, especially in comparison to the contemporary German left wing where vitalist ideas were also spreading. Thus, the early Frankfurt School embraced many aspects of the vitalist criticism of bourgeois democracy, as in Benjamin's or Adorno's positions. To some degree, vitalism united the two wings; Jünger was also influenced by left-wing positions and doctrines like the Soviet Five Year Plans, and his friend Ernst Niekisch, with his 'national-bolshevism', transgressed any easy categorization in terms of left and right. Horkheimer and Adorno's famous postwar *Dialektik der Aufklärung* remains deeply influenced by the vitalist criticism of Enlightenment.

5
Technology as Destiny in Cassirer and Heidegger

Continuing the Davos Debate

Hans Ruin

In recent years the legendary encounter and debate between Ernst Cassirer and Martin Heidegger in Davos in April 1929 has received a renewed interest, notably through the work of Michael Friedman, and also Peter Gordon.[1] It is then interpreted as a decisive event in twentieth-century philosophy, as an event both antedating and anticipating the sharp divides between different schools of thought that eventually came to characterize the philosophical landscape. At the time of the debate there was no clear and definitive division between an analytic-linguistic and a phenomenological philosophy, nor between a philosophy of culture in Cassirer's sense and an existential ontology. Nor had the political landscape taken on the disastrous shape that was to project many of the colleagues and discussants forever into different orbits, geographically and politically. In 1931 Rudolf Carnap – who was among the participants at the Davos meeting – published his sharp criticism of the inaugural address that Heidegger had delivered when taking over the Rickert-Husserl chair in Freiburg in 1928, thus establishing the fateful antagonism between logical positivism and existential phenomenology.[2] And from 1933 the political turmoil and Heidegger's initial support for the new regime, which included assuming for a time the rectorate in Freiburg, would forever colour the public image of his philosophy, and his relation to many Jewish colleagues.

Heidegger and Cassirer met only once after the Davos encounter. Apart from Cassirer's long review of the 1929 book on *Kant and the Problem of Metaphysics*, published in 1931, they did not further engage with each other's work, with the notable exception of Cassirer's critical remarks on Heidegger in *The Myth of the State*, completed just before his death in 1945.[3] Still it is possible to trace several philosophical itineraries along which one can see how they continue to work on related themes. There is a kind of virtual continued dialogue that deserves to be explored, not least for how it can help to throw light on their respective ways. A central theme in this respect is that of technology and the philosophical significance of the technologization of the world. When Cassirer published 'Form and Technology' in 1930 Heidegger had not yet published anything in this area, but in the years that followed it would emerge as a key theme in his thought, first presented in the public lecture 'The Age of the World Picture' ('Die Zeit des Weltbildes') from 1938 and culminating in the classic post-war essay 'The Question Concerning Technology' ('Die Frage nach der Technik') from 1953. My purpose here is to present a parallel reading of Cassirer's essay and some of Heidegger's contributions on this field, in order to show how Heidegger's work could be read partly as an implicit critical elaboration of Cassirer's approach, and how in some respects the stakes laid out in the Davos debate continue to determine also the way in which the problem of technology is treated in their respective works. Hopefully this reading can contribute to critically elucidating the sense and significance of both of their contributions. Before discussing in turn their writings, I will summarize some of the basic points of the Davos debate. I will also describe in broad terms the context of the philosophical concern with technology within which their interventions take place.

The Davos debate

At the time of the encounter in Davos, Cassirer and Heidegger were recognized as leading representatives for the two schools of thought which had competed for the initiative within German philosophy for several decades, neo-Kantianism and phenomenology. Originally they had both arisen from different environments within the manifold culture of neo-Kantianism. While Cassirer had his principal training with Hermann Cohen and Paul Natorp, Heidegger had written his doctorate on psychologism under Heinrich Rickert in Freiburg. It was only after receiving his doctorate that he turned his interest seriously towards Edmund Husserl and phenomenology, as it had been developed in the *Logical Investigations*. When Husserl replaced Rickert in 1916, Heidegger became his assistant. During the first part of the 1920s he put a lot of effort into developing Husserlian phenomenology in the direction of an existential ontology, partly in direct polemics against the neo-Kantians, as can be seen, for example, in the 1919 lecture course on the task and determination of philosophy.[4] This work culminated in the publication of *Being and Time* (*Sein und Zeit*) in 1927, in which he presents his ontological exploration of the meaning of being on the basis of a hermeneutical analytic of human existence, or *Dasein*.

In *Being and Time* Heidegger also insists on the necessity of undertaking a critical engagement (*Auseinandersetzung*) with tradition, a destruction of the history of ontology, not for the purpose of abandoning it, but rather in the service of releasing its inherent but repressed potential. To this destruction belonged, among other things, a new critical reading of Kant, which was also published separately in 1929 under the title *Kant and the Problem of Metaphysics*. In this work, the thinking of Kant is depicted as, in fact, on the way towards an ontology of human finite existence. The year before, Heidegger had also published an extensive review of the second volume of Cassirer's great work on the *Philosophy of Symbolic Forms*.[5] He speaks there in very positive terms of how not since Schelling has the mythical been treated in a more interesting way. At the same time, however, Cassirer is said to lack an ontological framework which would make it possible to do full justice to this phenomenon.[6] I will return shortly to this review, which anticipates in an interesting way the later critical development of Heidegger's thinking, as it comes to the fore, notably in the 1938 essay 'The Age of the World Picture'.

As Heidegger and Cassirer sat down beside each other to discuss their respective readings of Kant the issue was thus not simply a historical-philosophical argument. For in the question of who and what Kant actually was lay also the question of which one of these influential opponents was the most worthy inheritor to the best that German philosophy had brought forth. In other words, it was by no means only the past that was at stake, but the future of philosophy as such.

Cassirer opened the debate with a critical remark on the somewhat careless way in which the term 'neo-Kantianism' was used by Heidegger and others, to designate an integral school of thought.[7] In passing, he also suggested that Heidegger himself was more neo-Kantian than he chose to admit. From the different camps of the loosely connected school this was a remark that had been voiced before, notably by Natorp in his first review of Husserl's *Logical Investigations*. In many respects the neo-Kantianism of Natorp, as well as that of Wilhelm Windelband and later Rickert, could be seen as precursors and parallel efforts to that of phenomenology, both in terms of philosophical inheritance and in the concrete positions taken in such contemporary debates as psychologism and historicism, defending a transcendental philosophical approach in the face of different attempts at naturalization. Heidegger responded by giving a short sketch of the philosophical landscape as it had evolved from the middle of the nineteenth century. By that time, science appeared to have mapped out most of what there was to map out. Philosophy experienced a certain disorientation; what should now be its task? The response chosen by many philosophers was to try to re-establish, more or less explicitly in the spirit of Kant, philosophy as epistemology and critique of knowledge. Neo-Kantianism, as well as Husserl's early phenomenology, constituted in Heidegger's eyes precisely such a flight into theory of knowledge, away from the basic question of what is, and how it is – in other words: from ontology. But in the end this was also a betrayal of Kant. For Kant was not only trying to provide the natural sciences with an epistemology, he also sought to show the way towards the fundamental metaphysical and ontological questions. In this way Heidegger summarizes initially his own position, while also claiming to provide a more loyal continuation of the concerns of Kant.

In his response Cassirer returned to the lecture that Heidegger had given earlier, and to a theme which also occupies a central position in Heidegger's book on Kant, namely the doctrine of 'schematism'. It is a theme which on the surface would seem to belong to the periphery of the *Critique of Pure Reason*, but which in fact concerns something extremely important; namely how the general categories of understanding grasp and connect to the unique and manifold nature of exterior reality. In other words: how something like a uniquely human experience takes shape out of a combination of passive impressions and active concepts. At the centre of this process Kant places what he calls 'imagination', *Einbildungskraft*. It was in this imaginative capacity that Heidegger had sensed and sought to show that Kant too was on the way towards an understanding of the ontology of human existence. Also Cassirer confesses his great interest in this theme, while he insists that for Kant this did not constitute an end-point, but rather the beginning of a problematic. In the end the question concerns how a finite creature like man can reach insight into and knowledge of that which is not finite, in other words of supra-temporal, eternal truths, both theoretical and ethical. If one insists so strongly as Heidegger does, on the finitude of man and his existential being-towards-death, what happens then with reason and objectivity? Does Heidegger really want to renounce this objectivity? This is the profoundly resonating counter-question formulated by Cassirer.

In his subsequent response Heidegger turns the problem around by referring to Kant, and by showing how the question of the transcendent – that is, that which surpasses finite humanity – only becomes comprehensible in relation to human finitude, that it 'remains within the created and finite'. One misunderstands Kantian ethics, he says, if one only looks at the goal towards which it is directed, and does not consider the inner functioning of the law in relation to human existence. Not only ethics, but also ontology and the philosophical questioning itself, is based on a finite existence. Indeed, truth itself must be understood as emanating from human finite existence. For if there were no human being, there would be no truth. After these general declarations he directs his counter-questions to Cassirer: how does man enter into contact with the infinite? And how should philosophy relate to human anxiety? Should it seek to deliver him from it, or rather try to place it before him in a more definitive light?

In his response Cassirer recalls German Idealism and the words of Goethe: 'If you seek the infinite, explore the finite in all directions.' Precisely in this exchange the difference between them becomes most poignant. Despite the fact that their orientation and philosophical training were similar, it was two epochs and two different kinds of pathos that here stood over and against one another. While Cassirer cites Goethe, Heidegger implicitly takes his words from Kierkegaard and Nietzsche. While Cassirer represents a more confident critical idealism, Heidegger articulates the mood of the generation that had experienced the destruction of the old Europe in the trenches of the Great War – what could perhaps be described as an exposed, passionate nihilism, which at the same time seemed to promise something new and different. This dissonance in tone and orientation is highlighted even further in the review of Heidegger's *Kantbuch* that Cassirer published two years later in *Kant-Studien*.[8] Also in this text he his very respectful vis-à-vis Heidegger's interpretation, especially concerning the interpretation of schematism. But in his final remarks he argues that the entire interpretative horizon of Heidegger is determined by a Kierkegaardian conception of the world, which is fundamentally at odds with that of Kant, whose basic ethos is captured much better in Schiller's elevated poem 'The Ideal and Life' ('Das Ideal und das Leben'). He even goes so far as to portray Heidegger as an 'usurpator, who almost with armed force pushes his way into the Kantian system' ('als usurpator, der gleichsam mit waffengewalt in das kantische System hineindringt').[9] But it is important to see that this drastic and often-quoted image is embedded in a careful and in many ways respectful treatment of Heidegger's contribution to Kant scholarship.

The philosophical challenge of technology

After the Davos debate Heidegger and Cassirer met personally only once when Cassirer gave a talk in Freiburg. Cassirer published his review, and in his unpublished papers there is some more material concerning Heidegger; there is also the reference to Heidegger in *The Myth of the State*, as quoted above.[10] Heidegger occasionally refers to Cassirer in letters and in passing in a few texts, but after the review of the book on symbolic forms he does not address his work in writing. Still I believe that it is meaningful to look also for a more implicit continued dialogue with Cassirer on Heidegger's part, in which the principal issues voiced in the Davos debate continue to resonate in his work, but also in regard to themes that were not addressed on that specific occasion. One such theme, which unites them in their interests, is the question of technology, the impetus to interpret and understand philosophically the nature and meaning of technology as a destiny of man in modernity. Cassirer published his essay in 1930, at a time when Heidegger had not yet begun to work on this question. But in the years to come, it would move to the forefront also of his philosophical concerns, in lectures and manuscripts, the first published trace of which is the 1938 lecture 'The Age of the World Picture'.

My itinerary here is a critical reconstruction of the views on technology of Cassirer and Heidegger, in order to explore how Heidegger's thinking can be read as partly a direct response and elaboration of Cassirer's seminal essay. The value of such a comparison is twofold. First of all, and for the present volume most importantly, it can contribute to bringing Cassirer's standpoint to a heightened critical articulation. But by placing Heidegger's understanding of technology in the context of Cassirer's essay, it can also and likewise contribute to the clarification of Heidegger's philosophy of technology. In sum, we can see how, through the prisms of their respective efforts to think technology and technologization, a shared Kantian matrix is expanded and brought to bear on a new cultural and philosophical constellation.[11] Before continuing to the more specific interpretation of their respective contributions, I will try to sketch very briefly the cultural and philosophical situation in which both Cassirer and Heidegger take on this issue, a situation in which the question of technology with a new and sudden impact moves to the forefront of the philosophical agenda.

The question concerning technology is today the great issue of the future and destiny of mankind. It concerns what we can *do*, and what we are on the way to doing, unwillingly, through our very doing. In general, we can see how the philosophical question of technology has received ever-increasing attention and gained relevance throughout the whole postwar period, for very concrete reasons. The end of World War II marked the beginning of a new era, the atomic age, in which technical intelligence placed man in the role, if not of creator of the world, then at least as its potential destroyer, a threat to which recently has been added the awareness of artificially produced global warming. Through computer technology, artificial intelligence and through the discovery of the genome and its applications, man has entered the role of the mythical demiurge, in ways even the late nineteenth century could hardly even imagine.

In the philosophical discussion concerning the overall meaning and significance of this momentous transformation of man's relation to his surroundings, Heidegger's 1953 essay 'The Question Concerning Technology' has long upheld a canonical, if yet often disputed status. However, the emergence of a philosophical interest in technology is of an earlier date. In retrospect it seems that Germany in the 1920s was really a turning point in this respect. The first philosopher to take technology seriously was perhaps Karl Marx, who insisted on its world-shaping power. But it was not a full attempt to write a philosophy of technology. Two early such attempts were Ernst Kapp's *Grundlinien einer Philosphie der Technik* from 1877 and also Eberhard Zschimmer's *Philosophie der Technik* from 1914. But it is in the 1920s and early 1930s that the question of technology emerges rapidly as a defining issue of profound cultural-philosophical importance among the intellectuals at the time. An important book from this time, and a kind of stepping-stone for Cassirer when he wrote his essay, is Friedrich Dessauer's *Philosophie der Technik* from 1927, reprinted several times. In polemic against what he perceives as anti-technological sentiments in the present, Dessauer defends technology as also a spiritual (*geistig*) mission. Responding to a similar situation as the one analysed already by Marx, in which the modern organization of labour leads to alienation from production, he speaks in favour of a restored sense of the workers' community, not on socialist but on Christian grounds, in order to come to terms with the alienation of modern industrial labour.[12]

Dessauer was an important reference and a point of entrance to the whole problematic when Cassirer wrote his essay. The following year Oswald Spengler published *Der Mensch und die Technik*. With his characteristic blend of inconsistencies, bombastic phrases and cynical lucidness, he described technology as a natural extension of the will to power which animates the 'faustic culture'. According to his analysis, technology is at once the highest expression of man's will to power and a force that is enslaving man through the machine. Furthermore, it is leading to a devastation of the earth and its climate, and in the hands of the enemies of Europe it will lead to the continent's decline, if Europe has not already destroyed itself through an anti-technological *Schwärmerei*.[13] But civilization has itself become a machine in that it seeks to make everything machine-like. And in all of this Spengler sees a tragic destiny at work, a *Schicksal*. The following year Ernst Jünger published his visionary social-philosophical essay *Der Arbeiter: Herrschaft und Gestalt (The Worker: Dominion and Gestalt)*, in which he describes a new type of man, who through a higher fusion with technology is moving towards planetary domination.[14]

Heidegger does not officially participate in the first wave of increased interest in the philosophical significance of technology. But through the recently published volume 90 in his *Gesamtausgabe*, which contains most of his lectures and notes on Jünger and on *Die Arbeiter* in particular, from the years 1934–40, we can now fully appreciate the importance of Jünger's book in particular for the way the philosophical question concerning technology took hold of him also around this time, almost 20 years before the famous essay on technology. In sum, this brief overview gives a sense of how, during a few years, the question of technology moves in towards the centre of the cultural-philosophical debate.[15] In the context of this intensified intellectual preoccupation with technology, Cassirer's 'Form and Technology' ('Form und Technik') marks a significant contribution. In the following section I will give a comprehensive analysis of its content, with a specific view to the comparative argument.

Cassirer's 'Form and Technology'

Why, according to Cassirer, is technology a relevant and important theme for a general philosophy of culture? On one level, technology can seem easy to handle philosophically. If we pose the question 'What is technology or the technical?', we can point to the tool as an extension of the body whereby man expands his capacity for action and takes a firmer hold of his world. From this perspective, technology appears precisely as the natural extension of man's longing to control his surroundings and living conditions. But just as language transforms the world which it grasps, the tool, according to Cassirer, also leads to a change of meaning, a *Sinnwendung*.[16] It is this technologically induced change of meaning, or more precisely the influence of technology on the very meaning of experience and of the experienced world, that a philosophy of technology should explore. In this way of phrasing the problem we can detect what I referred to above as the 'Kantian matrix', expanded into a philosophy of culture. It is a philosophy of how the

world is symbolized and made meaningful through human spontaneous action. Technology is ultimately *a way of making sense*, or at least
this is the core of the philosophical problem of technology as Cassirer
understands it. In this way of focusing on the role of technology for the
creation and transformation of meaning he also anticipates a central
dimension of Heidegger's concern. For, as we shall see, it is precisely in
terms of how technology changes the meaning of nature and of being
that Heidegger, too, will address the problem.

Also in another principal respect, the overall orientation of Cassirer's
analysis anticipates that of Heidegger, namely in its historical orientation, even though it was precisely over the form and style of this historical analysis that Heidegger would diverge most clearly from Cassirer. For
Cassirer the question of the technological change of meaning requires
what we could call a mental-historical approach, or a history of men-
talities. His strategy is to explore, historically, the evolution of different
world images, *Weltbilder*. As the basic two different world images, he
points to the mythical and the technical. They respond, in turn, to two
figures or shapes of man, *homo divinans* and *homo faber*. In a polemic
with James George Frazer, who sees the magical as a kind of proto-
scientific exploration of causality, Cassirer insists on the great difference
between the magical and the religious world image, where the latter is
already influenced by the great transformative experience of technology. What technology brings to the magical, divinatory mentality is
an experience of willing and activity, over and against the experience
of a fundamental passivity with regard to nature. Through the active
use of technology man gets hold of nature as something separate from
himself, a distinct entity which he can influence through his will.
Through this blending of intervention and distance a new world image
emerges, a new step in the basic relation between man and his world.
The mythic-magical gives way to a discovery of nature as guided by a
norm and measure. It is only in and through this intervention that the
world can appear as *cosmos*, as order and form. This new image of the
world includes a new orientation of his vision, a *Blickrichtung*, through
which eventually also the fundamental a priori principle of causation
can appear, but only mediated through the use of the artefact, and the
experience of this use.

In this example we can sense clearly the character of Cassirer's development of Kant's transcendental idealism. To some extent we could say
that he takes Kant in a Hegelian direction, by historicizing the transcendental conditions of experience, showing how they become accessible
only through a historical process of experience and interaction with

nature. Out of this process, which on one level constitutes an accomplishment by man, man himself also emerges. In experiencing himself as an efficient cause, he also experiences 'a peculiar strengthening of his self-consciousness', and at the same time nature emerges as a 'modifiable, malleable material'.[17] Yet, in this promethean transformation, which on one level implies an upgrading of man's self-experience as a freely creative spiritual (*geistige*) power, there also seems to occur a parallel degradation of man as soulful (*seelische*) being. Cassirer is here recalling the distinction between these two dimensions of human consciousness, as developed in the contemporary cultural critic Ludwig Klages, where *Geist*, spirit, is the faculty of intellectual mastery manifested in outer deeds, whereas soul, *Seele*, is that aspect of man's self-awareness which is connected to the non-conceptual and affective interiority. He quotes Klages' analysis of how modern technical rationality as a 'vampiric' force feeds on and exploits the soul.[18] Even though Cassirer himself does not endorse such a depiction, nor its conclusion, he nevertheless urges his readers to take this modern widespread criticism of technical rationality seriously, and to try to understand the sense of the alienation from within the development of technology itself.

In order to do so it is important to see beyond the limit between an 'outer' and an 'inner' dimension of human accomplishments. For in every new worldly formation something of the inner is also disclosed. This in some sense Hegelian lesson is very important, not just for Cassirer's overall argument, but also for the comparison with Heidegger as we shall see later. Referring to the analysis of Ernst Kapp, Cassirer emphasizes that the work of technology, even though it is directed outwards, also and at the same time constitutes a means of self-knowledge. Man comes to know his inner self through his outward technical activity.[19] However, over the course of technical development man becomes increasingly estranged from the fruits and means of his own self-realization. Cassirer quotes Georg Simmel, who had spoken of it as a 'tragedy of modern culture', that the free spirit creates artefacts and structures in which it becomes increasingly entangled so as to experience them in the end as alien to itself.[20]

Cassirer himself does not follow the conclusions implied in these somewhat dystopian analyses concerning the cultural impact of technology. He is interested in understanding and explaining this widespread sense of alienation in modern life in its relation to technology. Most important perhaps is the description of how technology over the course of its perfection becomes more and more estranged from the original organic environment where it first emerged as the extension of the body. When craftsmanship is replaced by machine production something like the original rhythm of labour also disappears. The tool is no longer the extension of the arm, but the arm is rather the extension of the tool. The increasingly suffocating sense of technology as an outer force cannot be handled simply by pointing towards all the goods that technology brings with it; for even such a way of thinking still moves within the paradigm of a calculable utility. In the end the question is all about freedom and necessity. It is around this issue that everything is decided, according to Cassirer. If technology is not actualized as an idea of freedom, then 'it is condemned' (the literal German expression is a biblical allusion, that the 'stick is broken over technology', *die Stab über die Technik gebrochen*).[21] Technology must be thought of as a project of freedom in order for it to appear again as something free and liberating.

The problem, however, is that technology as such cannot set up its own goals. To its very essence belongs a goal-oriented rationality within a defined space of tasks, where it constitutes the best means towards a given goal. It cannot itself set the highest goals, which would require that it formulate an ethical task within an ethical community, a *sittliche Gemeinschaft*. Such a making-ethical (*ethisierung*) of technology he understands with reference to what Kant speaks of as an 'Ethico-Teleologie'. In Cassirer's view this requires a mobilization of new willpowers, through which technology should again be made into the servant and not the leader (a *Dienerin* and not a *Führerin*).[22] What he anticipates is a reinstallment of will, in which the motivation which once set the technical project in motion, but which somewhere on the way lost its orientation, through a new effort should overcome its alienation and take hold of itself and become master again of its situation. According to this figure of thought, which towards the end of the essay is put forth with a strong pathos, man is to take mastery also of himself. Technology, Cassirer writes, must not only vanquish nature, but the chaotic forces of the human being.[23] Behind all the negative aspects of technology which he has analysed up to this point, Cassirer sees this lack of fulfilment of its highest mission. Technology contains the potential of revealing to man his calling as a willing, ethical-rational creature; it has within it the potential of generating and shaping forms and idealities. But in order for this teleology to be fulfilled the human community must make this its explicit task.

If we were to try to summarize the historico-philosophical schema behind Cassirer's way of thinking about technology here we could depict it as follows: in and through technology the human mind makes a cultural leap into a new relation to nature and also in relation to itself. Technology changes the meaning of nature. Nature appears as something that man can control through his spiritual poietic activity, and consequently he will experience himself and his reason as an autonomous force in and over nature. But along the way towards this realization of his freedom, the human spirit will also experience the loss of its freedom, and come to view itself as passively exposed to technical and rational culture as to an alien force or destiny. But the way out of this alienated passivity in relation to his own creation does not go by means of its rejection, but, on the contrary, by entering deeper into its apparent destiny as, in fact, a latent freedom. In the final elated sentences of the text Cassirer even speaks of it as a 'community of destiny' (*Schicksalsgemeinschaft*) which is characterized precisely by freedom in serving, a freedom through servability, *Dienstbarkeit*, a formulation he takes directly from Dessauer's aforementioned book. By seeing and affirming, in a willing ethical effort, the technical as the most genuine manifestation of the will, man is to reappropriate his own historical situation. The more concrete political implications of this programme are not spelled out in this text. Presumably Cassirer is referring to this process both as an existential predicament to be realized by each and everyone, as a fulfilment of a personal free rationality, but also as a programme for a liberal and liberating politics, which takes a firm stance on its use and development of technology within the context of a social, ethical and political responsibility for the political community as a whole.

Before leaving Cassirer's analysis and turning to Heidegger, a few more points deserve to be mentioned here, and again for the benefit of the comparative discussion. Cassirer gives a very interesting analysis of the relation between science and technology, which anticipates Heidegger. Just like Heidegger later will do, he argues against the prevalent view, that technology should be seen as the extension of science. On the contrary, he writes (just like Heidegger 20 years later) that it is science in its modern experimental form that constitutes an extension of the technical approach to nature. It is not only the case that the creation of new technical instruments that extend the reach of the human senses enables a deeper exploration of nature. But it is rather the case that the technical-instrumental relation to nature is what grounds the scientific gaze. Indeed, he writes: the establishment of this way of visualizing nature is perhaps the greatest and most remarkable feat of technology.[24]

A second point, which I only mention here, is the brief but very important section in the text on the characteristic nature of *art* in regard to technology. In the spirit of Schiller, Cassirer insists on the specificity of the beauty of art, which is not the same as the beauty of a technological device, which is manifested in the optimal solution to a problem, in which the forces of nature are bound. In every poetical or plastic work of art, a new synthesis of subject and object is manifested, in an ideal balance of subjective expression (*Ausdrück*) and objective significance (*Bedeutung*). In this sense art preserves a particular and irreducible human ability, and even humanity as such.[25] This reference to *art* (and thus to the other sense of the original Greek *techne*) as a means to critically elucidate the nature of the technical is not really developed in the essay, but it points towards what will become decisive also in Heidegger's later analysis of modern technology, where art constitutes a space within which the bound instrumentality of technology can be critically reflected.

Heidegger's initial approach to the problem of technology

As briefly described above, Heidegger takes an early interest in the philosophical question of technology. Even though he refers very sparsely in his published writings to contemporary writers, with the exception of Ernst Jünger, it is clear from lecture notes and letters that he followed the debates of his time very closely, and that he was well read in the work of his contemporaries. I have no evidence that he actually read Cassirer's essay, and the argument here does not presuppose that he actually studied it. What is interesting to bring out in this context is the extent to which the general orientation of Cassirer's argument surfaces as a critical stepping-stone in Heidegger's own discourse, but also the extent to which he moves in parallel orbits around the central issues. In the now published lectures and notes we can follow the emergence of the theme of technology in his work. The first time he develops his analysis in public in a full manner is the Bremen lecture 1949, which was eventually published as 'The Question Concerning Technology' in 1953. But in the meantime he composed the important text 'The Age of the World Picture', first presented as a lecture at Freiburg University in 1938, in the context of a conference on the contemporary world picture.[26] I will present a reading of this text here which displays it as both an elaboration of Cassirer's question and also as marking a distinct philosophical criticism of some of the premises of Cassirer's analysis.

At the very outset of his 1938 lecture Heidegger states that what is of particular need of a philosophical reflection, *Besinnung*, is precisely 'science and machine technology'.[27] He then continues by making in principle the same point that had been emphasized by Cassirer concerning the relation between technology and science, namely that technology should not be seen as the outcome of modern science, but rather the contrary. However, he expands this argument somewhat in relation to that of Cassirer, even though it could be said to be a latent point also in the latter: that both science and machine technology are rooted in modern technology (*Technik*) as an overall metaphysics. Within this order, culture itself becomes an expression, a value and an artefact that can be organized within cultural politics. In other words, it implies a kind of general objectification of beings – what Cassirer would speak of as the objectification of nature. The proximity between their analyses comes out very clearly in the continued argument about science and technology, when Heidegger says that science – in its modern, that is, Cartesian-Galilean form – already presupposes that nature has been projected in a certain way, as the 'closed system of movement of spatio-temporal points of mass'.[28] This conception of nature is different from the one we find in the natural philosophy of the ancients. A decisive difference is the use of *experiments*. We should not understand experiments, Heidegger argues, as adding a level of certainty to science, but rather as peculiar activities which confirm the conception of nature as a closed and calculable system of movement. In other words, a technical understanding of nature is already present in the very experimental method of research. Consequently, the scientist who seeks truth is in some ways already a technician, who produces results within a defined order or establishment, an *Einrichtung*.[29] From this very general perspective on scientific activity, Heidegger does not see an essential difference between science and the humanities, since they both share a common approach to their objects, which comes down to 'representing' that which has permanence and existence. Within this system, research will inevitably obtain more and more the character of a profession, a *Betrieb*. In very broad strokes Heidegger evokes the emergence of modern society with its industrial organization of result-oriented research, in which the individual researcher becomes increasingly like a technician who performs already defined tasks. In Heidegger's interpretation, this transformation and development rests on a metaphysical transformation in the way knowledge itself is conceptualized and understood. As the defining moment of this new conception he points to Descartes. In Cartesian metaphysics we can sense how the whole world is projected as a representation, and the subject of knowledge as the ground of knowledge.[30] In the general description of a transformation of the relation between man and nature, the analysis resembles to a surprising extent that of Cassirer's, even though Cassirer would seem to date this 'promethean' moment much earlier than Cartesian modernity (even though the chronology of the historical development is more fleeting in his essay).

The resemblance may not strike one at first, since the pitch of the analysis is different. Cassirer seems to want to write the history of a transformation of the system of knowledge and consciousness, whereas Heidegger seems to have a more open, critical perspective on this modernity. Yet, we should not overstate the critical dimension of Heidegger's engagement, since it has often been misunderstood as anti-modernist. The point of his analysis is not primarily to criticize, for example, Descartes for his metaphysical invention, but rather to bring to philosophical, reflexive awareness, a transformation which guides inadvertently the way that modern scientific knowledge projects and constructs its objects. Just like in the analysis of Cassirer, Heidegger is working within a historicized Kantian matrix, trying to come to terms with the role and significance of technology within the system of knowing as such. Yet, in his more poignant critical remarks on the projecting, representing consciousness, Heidegger also implicates the subsequent tradition of critical and transcendental idealism, which for Cassirer still functions as a guiding ideal. It is in this way that the debate on Kant continues to structure and to guide even more strongly the discrepancies between their respective analyses. For in the years that passed between the publication of *Being and Time* (1927) and the book on Kant (1929), Heidegger had performed a monumental critical confrontation with the tradition of German idealism, notably the work of Kant and Schelling, and what he saw as its most radical extension, Nietzsche's philosophy of a will to power.[31]

In the course of this work, he excavated a persistent tendency towards making being the object of a representing reason, and to an image, a *Vorstellung* and a *Bild*. That he addressed the specific theme of *Bild*, image, in the actual context had to do first of all with the title of the symposium within which it was presented, namely 'The Grounding of the World Picture of Modernity'. But in his critical elaboration of this theme we can also see how he picked up again a thread from the review of Cassirer's *Philosophy of Symbolic Forms* ten years earlier, concerning precisely the significance of the fact that the world is projected as image, as *Weltbild*. When we ask a question about world images we are not just sharing an innocent concern, we are also contributing to the creation of the world as image. A *Weltbild*, he wrote, understood essentially does not signify an image of the world, but the world understood as image.[32] Already, in seeking the answer to the question concerning the nature of our modern world image, we have committed ourselves to a decision about being in its totality. For in this question we seek, and find, the being of beings in the representativeness, *Voregestelltheit*, of beings. It is precisely this thinking about the world as image, as world image and also as worldview, which is mirrored also in planetary technology. For in the extension of this technical development all distances are constantly shrinking with the increase of speed. When we critically reflect on technology we therefore also critically engage with how the world is conceived as world image.

I am here shortening Heidegger's more complex argument down to what I see as its basic structure, in order to develop the critical comparison. A conclusion to be drawn from this reading of the two texts is that, to a certain extent, Heidegger and Cassirer approached the philosophical problem of technology in surprisingly similar ways, which can be understood in terms of a shared Kantian matrix. In technology they refused to see only the extension of man's desire for control over nature. Technology is also, and more importantly, a system for representing nature and for organizing the relation between consciousness and nature, between subject and object. Furthermore, they both shared a concern with a certain alienating effect of this technological matrix for the understanding both of nature and of mind, which called for a philosophical analysis. But it was also in the formulation of how philosophy should confront the challenge of technology that they chose separate paths. While Cassirer reached back towards a Kantian practical-teleological rationality, which should recapture the element of freedom in technology itself as a cultural value to be projected anew, Heidegger pointed towards that shaded zone out of which planning, calculating, representing and image-making thinking emerges in its own utter incalculability, 'its invisible shade'. This shaded background will not itself become visible if we try to escape modernity towards a lost past, nor through a simple negation of what it stands for. Instead, the role of a philosophical (*besinnende*) reflection is to make the very situation appear in its facticity and unicity. In other words, we can make it visible, we can represent it to ourselves, so as to think and discuss it, but in the end we cannot hope to master this predicament, least of all by projecting it as a logical sequel in a development of world images. For in doing so, we simply reinstall its inherent logic.

In the 1938 lecture, Heidegger is quite brief when it comes to the meaning of this peculiar philosophical reflection or *Besinnung*. In many of the later texts he would elaborate it further. Still the contrast, both subtle and sharp, between his and Cassirer's approach can be articulated on the basis of this first outline. For Heidegger, the solution to the philosophical-existential problem that technology poses to us cannot consist in taking on, in a firm decision of the will, the challenge of articulating its latent value. And the reason for this is that such a decision will only risk reduplicating the initial problem, since it moves in the direction which technology has already anticipated. Already, in his review of Cassirer's principal work, Heidegger had questioned the use of *Bild* and *Gebilde* as a premise for interpreting the relation between the mythical and the logical. And he had concluded that it is precisely in the extension of Kant's Copernican revolution that reality can become the creation (*Gebilde*) of an actively shaping consciousness.[33] In a similar critical spirit he wrote in the published footnotes to the 1938 lecture, that the modern reference to values (*Werte*) was in itself a symptom of a certain way of comporting oneself vis-à-vis reality. In the age of the world picture, being becomes a value, becomes 'cultural values, that become the expression of the highest goal of creation in the service of securing man as subject'.[34] The target of this critique is not specified. To some extent it highlights certain aspects of the National-Socialist cultural politics, and its cult of technology and progress, in ways which he began to develop more explicitly in his unpublished works from around the same time, notably the volume entitled precisely *Besinnung*.[35] But from the continued argument in the notes, which takes up in a critical spirit the development of German Idealism, it is clear that he saw it as applicable to a larger philosophical situation. For our part, we can note the obvious proximity to some of the formulations in Cassirer's essay. Heidegger would undoubtedly have seen in Cassirer's suggested solution precisely this risk of reproducing the problem, by trying to mend the situation by setting up more firmly a set of orienting values. The question is not, as I see it, if he would have come to this conclusion had he addressed explicitly the argument of Cassirer in the technology essay, for they are already implied in his earlier criticism of his work. What is of philosophical significance is the more exact relevance of these reflections, and the extent to which they can genuinely be said to address Cassirer's analysis. But before trying to answer these questions, we need to look also at how Heidegger develops his understanding of technology in the subsequent years, and in particular in his most famous contribution 'The Question Concerning Technology' from 1953. After reading this text we are in a better position to develop, in a final section, a tentative dialectical conclusion in response to the questions raised.

Heidegger's later approach to the question of technology

During a five-year period following the end of the war, Heidegger was forbidden to teach as a result of his activities as rector in 1933. During this time, he composed several important texts, including the 'Letter on Humanism' from 1947. It was also during this time that he gathered his thoughts on technology, as outlined in 'The Age of the World Picture', but which in fact dates back to unpublished manuscripts from the mid-1930s. In 1949 he gave four talks at a cultural club in Bremen, one with the title 'The En-framing' ('Das Ge-stell'), which, in a revised version, was later presented as a lecture at the School of Polytechnics in Munich in 1953 under the title 'The Question Concerning Technology'.[36] It is a text which is not easily interpreted, and which demands a careful reading in order to fully make sense. Here I can only bring out some of its most important points and movements, in order to develop the overall argument concerning the comparison with Cassirer.

To begin with it is important to keep in mind the idea formulated already in the lecture on 'The Age of the World Picture', namely that when we try to see and conceptually come to terms with a certain phenomenon we also have to pay close attention to *how* we approach it, that we take into consideration the method in the sense of the way towards the matter at hand. For there will always be a risk that we let ourselves be guided by a thought model which in the end makes us blind precisely to the phenomenon which we are trying to interpret and understand. This remark is absolutely decisive for understanding how Heidegger, in the later text, approached the question of technology, namely as a question which must – as he says – aim at 'building a path in language in the direction of technology', in order for its essence to reveal itself to us. It is only when the essence of something is disclosed to our existence that we can establish something like a 'free relation' to it. In other words we should not satisfy ourselves with looking for the appropriate philosophical theory about technology and the technical, but we must somehow accomplish a practical transformation of thinking itself, so as to make it attentive to the essence of the technical. Through such a praxis something like a *freer* relation to technology should be possible, not by radicalizing technology, not through a willing acceptance and control of the technical, but rather by becoming attuned to the technical as a sort of destiny to which we always already belong. For, in Heidegger's understanding, technology constitutes a destiny, the destiny of metaphysics and of modern rationality, in its will to dominate and seek control of the world and of being as image, as representation. The premise for the whole argument is thus that we, in our customary way of thinking, are already chained to technology, even when we believe ourselves to be most free from it.

In his continued attempt to give voice to this essence of technology, Heidegger dispels again the standard instrumental definition of it as simply a tool. In the end technology must be understood as a way of 'making true' or of 'revealing' a world.[37] This at first surprising – but, for the continued argument, decisive – formulation can be understood if we recall the phenomenological and transcendental-philosophical of Heidegger's thinking file. Technology is a way through which nature manifests itself in its being. Cassirer, and also Husserl, would have expressed it differently, in terms rather of how it accomplishes a new 'meaning', a change of meaning, a *Sinnwendung*. Heidegger insisted that this really concerned the *being* of nature, but the idea is, in fact, not so far from the one expressed by Cassirer concerning the transformation of the appearance or phenomenon of nature in and through technology.

How then, does technology disclose nature? As something to challenge or exploit, *Herausfordern*; the kind of disclosure latent in technology is precisely such an exploiting challenge.[38] The idea has its parallel in Cassirer's reflection that the greatest effect of technology is to transform nature into 'a region of free possibilities'.[39] In Heidegger this means that through the very technical comportment nature appears as something positioned and placed, as a 'standing reserve' (*Bestand*). The technical comportment reaches us as a demand to think and act in its direction. It is this sense of a demand which at the same time discloses nature as something to challenge and exploit, which he summarizes under the neologism *Gestell*, often translated as 'framing' or 'enframing' – a term in which he asks his readers to perceive also the echo of placing (*Stellen*), demanding (*Bestellen*) and representing (*Vorstellen*).[40] In a famous example he describes the water plant in Rhine as also transforming the river into a resource built into the plant. The being of the river becomes a different one through technology. But it is important to see that this demand is not something which happens to nature as a result of man's voluntary activities. It manifests itself as a demand directed to and from within man. In the *Gestell* man is called and challenged to challenge nature.[41]

In Heidegger's view, science in its modern form is mostly already an integrated part of the *Gestell*, as a name for the essence of technology. Just as we could extrapolate a kind of basic historico-philosophical schema from Cassirer's analysis, we can in a similar way detect an overall scheme in Heidegger's description. Technology is inextricably tied to the metaphysical impulse to determine the being of beings as something representable, as substance, force, or as will. But this impulse will only appear as such – that is, as an impulse – if we approach it from the position of a reflexive, *besinnende*, thinking which permits it to resonate as such. In such a listening attention we can become attentive to it as something which characterizes also ourselves in our urge to dominate it conceptually. This is when we permit the essence of technology to truly resonate, and thus to open up a more free relation to it. We cannot escape technology, but neither should we hope to become its masters, because in the very ambition to achieve mastery, we are still thinking and operating within the reach and scope of the technical.

Concluding comparative remarks

To what extent should we imagine that Cassirer could have followed Heidegger in this later analysis? Cassirer also saw technology as structuring the very way in which man confronts and conceptually grasps nature. He, too, conceived of technology a destiny, a *Schicksal*, in the sense that its force has always already taken hold of human thinking and reshaped its relation to nature, leaving us with a responsibility to think and reflect on a situation from which we cannot step back, and behind which we can no longer reach. Its effect has, in this sense, always already taken place. For both of them the issue is to articulate what it could mean to establish a freer relation to technology. They both remain oriented by a Kantian imperative according to which philosophy is a work of freedom, and ultimately a work to liberate thinking itself. The decisive and critical difference between them, however, manifests itself in the different suggestions as to how we should respond philosophically to this predicament. It manifests itself in the very mode that philosophical, reflective reason should respond to that technical-instrumental rationality of which it itself is to some extent always already a part. Whereas in Cassirer the response is one of mobilizing a latent ethical-teleological rationality, Heidegger's analysis casts a doubt precisely on such a solution. In his philosophical therapy the challenge becomes rather that of becoming more attentive and critically aware of the specific nature of the 'call' of technology itself. The challenge is to be able to hear this call while not being controlled by its inherent urge for mastery, and only thus to hear it in a liberating way. The premise for this therapeutic argument is that technology is also a mode of thinking which exerts a measure of control over us and the way we normally present and represent beings. Thus, freeing ourselves from the controlling dimension of the call of technology ultimately also means freeing ourselves from a certain aspect of ourselves.[42]

It would not be enough, in Heidegger's view, to recapitulate the world image or worldview within which we stand, and then try to reform it with the help of a more rational ethics. Instead we must move beyond the temptation to grasp the world as image, and try instead to think and experience it from the finite horizon within which we find ourselves. In relation to technology this means that we cannot aspire to step outside the technical simply by grasping its essence. What we can hope for is to experience it as determining the finite situation in which we stand, and as an aspect of how we exist.

In a critical comparison of Cassirer and Heidegger, Wayne Cristuado recalls Heidegger's famous dictum that 'only a god can save us'. He then argues that Cassirer's version of critical idealism, with its continued belief in some sense of autonomy, would set him distinctly apart from the philosophical fatalism implied by this remark.[43] Cristuado echoes the analysis of Habermas, that the later Heidegger's renunciation of willing was an effect of being crushed by the actualities, and that his thinking developed into a kind of fatalism, which gave up all hope of transforming the present. It also recalls Cassirer's own final criticism from 1945 in *The Myth of the State*, as quoted earlier, which accused Heidegger of philosophical fatalism. Despite its apparent pertinence, I consider this common criticism to be misleading as a reading of Heidegger's project. First of all, it fails entirely to account for his prevailing preoccupation with the problem of actually seeking an articulation of what it could mean to think *freely* in the age of technology. It also fails to account for his acute attentiveness to the internal controlling forces of technical rationality itself, related in some respects to what Adorno and Horkheimer would also explore in the *Dialectics of Enlightenment*, and that would also become a central theme in the work of Foucault. The point of Heidegger's therapy, as vague as it may seem, is not to give up and simply abide, but to actively seek and cultivate a new relation to the essence of the technical, not to submit oneself to its power.

If we return again to where we began, in the Davos discussion, we can try to imagine what and how Cassirer would respond to such a suggested therapy. Let us suppose that Heidegger's point is that we cannot hope to transcend entirely the finite historical space of technology, for it marks a thrownness and destiny within which we stand, and every attempt to project it along an ideality or a universal value will therefore just reproduce inadvertently that very same finitude. Then Cassirer would probably respond that the very idea of such a closed historical destiny already presupposes some sort of ideality and a universality in order to be understood as such – in other words, some sense of rationality. And in this argument, as he also stated it in the Davos debate, he has of course a strong dialectical point. But Heidegger's response would probably be that even though every notion of finitude logically implies that of the infinite as its defining opposite, we should not believe that we can grasp the world from this realized pole of infinity. We can perhaps experience its evanescent nature in our attempt to make our situation comprehensible, but if we believe that we can grasp it, we will end up in an illusory mastery, in which we are in the end controlled by that which we sought so control. The idea that technology could be mastered and rationally guided by means of an effort of the will would, in Heidegger's understanding, place us in such a trap of illusory self-guidance.

In saying this he would still not be urging us to give up and wait for a saviour, but to remain wakeful to what this call for control carries within it. No doubt Cassirer would call into question such a strategy for giving way, nevertheless, to a kind of fatalism. And in the reference to technology as an inescapable destiny, he would hear the echo of a proto-totalitarian discourse which he had learned bitterly to be attentive to in the realm of politics. If we do not take charge of our situation, if we do not at least try to formulate to the best of our knowledge the ethical and political goals with which we would like to see technology develop, we leave its application in the hands of future despots or simply a random rule. The enormous political efforts to establish a global consensus on the fabrication and proliferation of nuclear arms, and also of the civil use of nuclear energy, could be seen as examples of both the necessity and the at least partial success of such an approach to the threatening reality of modern technology.

Heidegger would most likely respond that the approach to technology for which he had argued should not be seen as contradicting any efforts to make current and future technology more safe and more useful to humanity. But he would strongly insist that if we do not try to genuinely understand the essence of the technical, and thus also the essence of the kind of rationality of which our metaphysical-philosophical tradition is partly an outcome, we will remain blind to an inherent totalitarian and controlling temptation which lies at the root of this technical rationality itself. In other words, technology is not a neutral force, simply to be used according to the application of an ethical imperative. Its reach over human thinking and experience is greater than that. And only if we permit ourselves to contemplate it from the viewpoint of a certain disengagement can we understand the nature of this imperative. If we, on the other hand, refrain from this level or mode of reflection we risk losing something very essential to philosophy itself. For philosophy, in the age of technology, will easily fall prey to the technological imperative, in which its mode of questioning gives way to a technical-scientific rationality, where cybernetics, cognitive science and applied ethics will appear as the logical outcome of its tradition. This is why an inherited sense of rationality and universal explicability must continue to be reflected from within an experience of the finitude of our historical-destinal situation. Rationality itself, not least in its technical-instrumental application, is in itself a finite destiny, a space of meaning and understanding within which we stand, but which we can only grasp as such as long as we do not project ourselves inadvertently in its direction.

To what extent would Cassirer have followed him also in this last argument? Here the issue is not the interpretation of Kant and his critical idealism, but the very possibility of a rational ethical critique of the technological present. We will never know. But, just as in the case of the Davos debate on the inheritance of Kant, we can see how their respective standpoints on the question of technology together delineate a space of questioning which the present has by no means transcended, but which, on the contrary, constitutes a prevailing situation within which we stand.

Notes

1. Michael Friedman, *A Parting of the Ways: Carnap, Cassirer, Heidegger* (Open Court: Chicago, 2000), esp. chapter 1. The present chapter was written and submitted before the publication of Peter Gordon's *Continental Divide: Heidegger, Cassirer, Davos* (Cambridge: Harvard University Press, 2010), which is why it has no further references to this excellent study. Gordon's book greatly expands the contextual understanding of their encounter and its consequences for twentieth-century philosophy, but it does not address specifically the question of technology. The present chapter was written in the context of a research project 'Nihilism, Lifeworld, Technology', which was generously supported by the Baltic Sea Foundation. I was first motivated to write on the particular theme of Cassirer and Heidegger as a result of an invitation by Bernd Henningsen and John Michael Krois to Humboldt University in Berlin in January 2008. I have since then also presented versions of the text at Helsinki Advanced Collegium, with Sara Heinämaa, and with prepared comments by Fredrik Westerlund; and also at Gothenburg University, following an invitation from Mats Rosengren. I am grateful for the remarks and criticism I received during these occasions, and also for the detailed and illuminating comments from the editors of the present volume.

2. 'Überwindung der Metaphysik durch logische Analyse der Sprache', *Erkenntnis* 2 (1931). In English translation 'The Overcoming of Metaphysics through Logical Analysis of Language', in Michael Murray (ed.), *Heidegger and Modern Philosophy* (New Haven: Yale University Press, 1978), 23–34.

3. Ernst Cassirer, *The Myth of the State* (New Haven: Yale University Press, 1946), 292f. Against the background of a critical examination of mythical elements in contemporary, authoritarian political philosophies, Cassirer here sharply distinguishes Heidegger's 'Existenzialphilosophie' from its Husserlian origin, pointing in particular to what he holds to be its refusal to recognize something like 'eternal' truths. The remark is misleading as a reading of *Sein und Zeit*, which is a work that explicitly explores an existential a priori, but it is still pertinent, not least in the light of what was the central theme already in their Davos debate – namely the status and standing of the ideal and universal. Even though Cassirer does not accuse Heidegger's thinking of having 'any direct bearing on the development of political ideas in Germany', he still suggests rather sharply that it did 'enfeeble and slowly undermine the forces that could have resisted the modern political myths'. One reason for this conclusion is that he sees in the thought of Heidegger and others in his generation a 'return of fatalism'. This last remark is of particular importance for the argument in the present chapter, and I return to it in the concluding discussion. See also note 10 below for Cassirer's remarks on Heidegger.

4. Martin Heidegger, *Gesamtausgabe*, vol. 56/7: *Zur Bestimmung der Philosophie*, ed. Bernd Heimbüchel (Frankfurt am Main: Klostermann, 1987).

5. Martin Heidegger, 'Ernst Cassirer: Philosophie der symbolischen Formen. 2. Teil: Das Mytische Denken. Berlin 1925', first published in *Deutsche Literaturzeitung* 21 (1928), 1000–12; reprinted in *Kant und das Problem der Metaphysik* (Frankfurt am Main: Klostermann, 1977/91), 255–70.

6. This ambiguous reception, at once appreciative but also signalling an insufficient ontological grounding, almost literally recalls his formulations from the review of Jasper's *Psychologie der Weltanschauungen*. See Martin Heidegger, 'Anmerkungen zu Karl Jaspers "Psychologie der Weltanschauungen" (1919/21)', in *Wegmarken* (Frankfurt am Main: Klostermann, 1976), 1–44.

7. The protocol of the whole debate was printed first as an appendix to Heidegger's *Kant und das Problem der Metaphysik* (Frankfurt am Main: Klostermann, 1976/91), 274–96.

8. Ernst Cassirer, 'Kant und das Problem der Metaphysik: Bemerkungen zu Martin Heideggers Kantinterpretation', first published in *Kant-Studien* 36 (1931), 1–26); reprinted in Ernst Cassirer, *Gesammelte Werke*, vol. 17: *Aufsätze und kleine Schriften* (1927–1931), ed. Tobias Berben (Hamburg: Meiner, 2004), 221–50.

9. Cassirer, 'Kant und das Problem der Metaphysik: Bemerkungen zu Martin Heideggers Kantinterpretation', 240.

10. See an analysis of these remarks in John Michael Krois, 'Cassirer's Unpublished Critique of Heidegger', *Philosophy and Rhetoric*, 16:3 (1983): 147–59. Krois here also suggest that one reason why Heidegger did not finish his planned review of the last volume of Cassirer's great work could have been that he found no way of mapping its conclusions onto the standard image of neo-Kantianism.

11. Before Michael Friedman's aforementioned book (see note 1) there was not much work on Cassirer and Heidegger. But following Peter Gordon's study

(see note 1) the interest in this constellation is growing steadily. There are some previous shorter analyses of the Davos encounter, which include more general assessments of their respective positions, especially Wayne Cristuado's 'Heidegger and Cassirer: Being, Knowing and Politics', *Kant-Studien*, 82:4 (1991), 469–83, to which I will return briefly towards the end. I am grateful to Aud Sissel Hoel for pointing out these sources.

12. Dessauer also directs himself to the technicians, whom he sees as a decisive but alienated force in society, who are never taught to see technology in its cultural context. A telling quotation in this respect is the following (in my translation): 'Up until recently an engineer, after having completed his studies, had not heard a single word about what technology really is, from out of where it springs, and how it is intertwined with the nature of man and the cosmos, nor about its utmost goal and its limits, how it is connected with the capacity for knowing, the imagination and the moral law within us. One is taught how to handle machines, but not that there is also a human being beside it.' Friedrich Dessauer, *Philosophie der Technik: Das Problem der Realisierung* (Bonn: Verlag Friedrich Cohen, 1927/33), 17. Another contemporary writer on the need for a more spiritual interpretation of technology, who is also an important reference for Cassirer, was Max Eyth, who had published the book *Lebendige Kräfte: Sieben Vorträge aus dem Gebiet der Technik* (Berlin: Springer,1924).

13. Oswald Spengler, *Der Mensch und die Technik: Beitrag zu einer Philosophie des Lebens* (Munich: Becksche, 1931).

14. Ernst Jünger, *Der Arbeiter: Herrschaft und Gestalt* (Stuttgart: Ernst Klett, 1932/81). Surprisingly, this seminal text was never fully translated into English.

15. There are already a few studies devoted to different aspects of this philosophical territory, notably Michael Zimmermann's *Heidegger's Confrontation with Modernity* (Bloomington: Indiana University Press, 1990), a book which for its historical context builds much on the somewhat earlier *Reactionary Modernism* by Jeffrey Hersch (Cambridge: Cambridge University Press, 1984).

16. See Ernst Cassirer's 'Form and Technology' in this volume, 25. See also 'Form und Technik' in Ernst Cassirer, *Gesammelte Werke* [hereafter ECW], vol. 17: *Aufsätze und kleine Schriften* (1927–1931), ed. Tobias Berben (Hamburg: Meiner, 2004), 139–83, 151.

17. Ernst Cassirer, 'Form and Technology', 34/ECW 17, 163.

18. *Ibid.*, 35/ECW 17, 165. For a similar witness of the contemporary reaction against technological rationality, he also refers, with more sympathy, to Walter Rathenau, who had spoken of modernity as a restless exploitation, and as meaningless consumerism. *Ibid.*, 48/ECW 17, 181.

19. *Ibid.*, 38/ECW 17, 168.

20. *Ibid.*, 41/ECW 17, 172.

21. *Ibid.*, 41/ECW 17, 173.

22. *Ibid.*, 49/ECW 17, 182.

23. *Ibid.*, 49/ECW 17, 183.

24. *Ibid.*, 44/ECW 17, 175.

25. *Ibid.*, 47/ECW 17, 180.

26. Published in the collection *Holzwege* (Frankfurt am Main: Klostermann, 1950/80), 73–110; also in English translation by William Lovitt, 'The Age of

the World Picture', in *The Question Concerning Technology and Other Essays* (New York: Harper & Row, 1977).

27. The text was published in the collection *Holzwege* (Frankfurt am Main: Klostermann 1950/80), 73–110. For a more detailed discussion of the sense and translation of *Besinnung*, see my 'Prudence, Passion, and Freedom: On Heidegger's Ideal of Besinnung', *Giornale di Metafisica* XXVIII (2006), 29–52.

28. Martin Heidegger, 'Die Zeit des Weltbildes', *Holzwege* (Frankfurt am Main: Klostermann, 1950/80), 76 (my translation).

29. Heidegger, 'Die Zeit des Weltbildes', 83.

30. *Ibid.*, 85, where he writes of how 'being becomes the object of a representation', and truth is understood as certitude of such a representation, with man as its subject.

31. For a more detailed reading of Heidegger's confrontation with Kant and Schelling, primarily on the role and meaning of freedom, see my 'The Destiny of Freedom – In Heidegger", in *Continental Philosophy Review*, 41:3 (2008), 277–99.

32. Heidegger, 'Die Zeit des Weltbildes', 87.

33. Heidegger, 'Ernst Cassirer: Philosophie der symbolischen Formen', 265.

34. Heidegger, 'Die Zeit des Weltbildes', 99.

35. Martin Heidegger, *Gesamtausgabe*, vol. 66: *Besinnung* (Frankfurt am Main: Klostermann, 1997). Together with the major work *Beiträge zur Philosophie*, also published posthumously in 1989 (*Gesamtausgabe*, vol. 65), this book was composed for an unknown future at a time when it had no prospect of being published. For his criticism of Nazi cultural and scientific politics, see in particular 28, 122 and 173f.

36. The essay is included in the volume *Vorträge und Aufsätze* (Pfullingen: Neske, 1954). In English translation 'The Question Concerning Technology', in D. Farrell Krell (ed.), *Basic Writings*(London: Routledge, 1978/93), 307–42.

37. *Ibid.*, 16/319.

38. *Ibid.*, 18/320.

39. Cassirer, 'Form and Technology', 44/ECW 17, 176.

40. For a more detailed analysis of this particular notion and its role in Heidegger's thought, see my 'En-framing', in Bret W. Davis (ed.), *Martin Heidegger: Key Concepts* (Chesham: Acumen, 2010).

41. Heidegger, 'Die Frage nach der Technik', 21.

42. In this way of phrasing the problem and the situation we can hear the resonance of Heidegger's reading of Meister Eckhart, and his call for a *Gelassenheit* in regard to oneself and to the entities that demand our attention. See in particular Heidegger's *Gelassenheit* (Pfullingen: Neske, 1959), where the relation to Eckhart is also briefly addressed. This critical elaboration of the will is also the dimension of Heidegger's thought which opens it most visibly to certain traditions of Asian thought. For a fine analysis of this problem in a comparative context, see Bret W. Davis, *Heidegger and the Will: On the Way to Gelassenheit* (Evanston: Northwestern University Press, 2007).

43. Cristuado, 'Heidegger and Cassirer: Being, Knowing and Politics', 483.

6
Technical Activity as a Symbolic Form

Comparing Money and Language[1]

Jean Lassègue

My aim in this article is to discuss Ernst Cassirer's conception of technical activity in relation with the notion of a symbolic form. The questions I shall raise are: in what sense is technical activity a symbolic form? What kind of relationship does technical activity entertain with other forms and in particular with language? And what does this relationship teach us about technical activity today? After trying to answer these questions in a theoretical way, I shall use the comparison between money (interpreted as a technology) and language (interpreted as the paradigmatic symbolic form) as a case study.

A disparity in Cassirer's philosophy

Linguistic versus technical activity

In Cassirer's *Philosophy of Symbolic Forms* (3 volumes), the relationship between symbolic forms is always interpreted from the point of view of language, since it is ultimately the human attitude towards language which makes new symbolic forms possible: the vision of the world remains mythical, says Cassirer, when the human attitude towards natural language is that of proximity, and it becomes scientific only when the human attitude towards natural language is that of a divorce. If mythical thought is considered as the most primitive symbolic form in the historical order, it is nevertheless the attitude towards language which allows for the further symbolic developments of humanity. In the case of technical activity, how should its relationship towards myth and language be described? Is technical activity opposed to myth and language, the same way science is? Or, on the contrary, does technical activity possess a common ground with myth and language?

If we take into account Cassirer's philosophy of symbolic forms only, we are left with speculations, since technical activity is not analysed as such. Consequently, technical activity is not considered as a primary symbolic form alongside language, mythical thought or science and we do not even know if it is a symbolic form at all. Rather, what is stressed in the first two volumes, as well as in other works of the same period,[2] is the proximity of mythological thought to language. In the philosophy of symbolic forms in general, the relationship between language, on the one hand, and mythological thought and science, on the other, is always kept in balance: the most mythical layers of meaning are always active in language, which is nonetheless a means of escaping from these very layers at some later step in the development of symbolic forms. Language possesses, therefore, the unique versatile power of pervading all symbolic forms, from the most mythical to the most scientific ones. Thus, at the time he developed his philosophy of symbolic forms, Cassirer's philosophical agenda was clearly focused on the attitude towards language, since it is this attitude which directly modifies the inner balance between other symbolic forms, especially mythical and scientific thought. Hence the crucial role played by language right from the beginning, from the first volume and onwards in the other two.

The importance of 'Form and Technology' ('Form und Technik'), published in 1930,[3] clearly derives from its focusing on the question of the nature of technical activity in its relationship with the notion of symbolic form. What is striking in this article is that an opposition is being made between language and technical activity, on the one hand, and mythological thought, on the other: when man was still immersedin a mythical vision of the world, says Cassirer quoting Herder, language and tools opened up an entirely new realm by severing him from the immediate present and attracting him into the workings of mediation. But one has to ask the following question: is the linguistic mediation of the same kind as the technical one? In the case of language, two dimensions seem to be at stake: the dimension of *expressivity* – that is, the fact that one has to acknowledge that he or she is the addressee of what happens in a particular situation – and the dimension of *semiosis* – that is, the fact that any given meaning of a sign is appointed by repetition, and it is thanks to other signs that a sign is recognised as such, otherwise the sign loses its character as a sign and becomes a mere piece of matter deprived of meaning. In the case of technical activity, the mediation is not of the same kind because it operates between a tool and *Nature*, defined as this entity which neither addresses anything to humans nor depends on their semiotic conventions: hence the obvious discrepancy for any unbiased reader between the two kinds of mediation. Nevertheless, in 'Form and Technology', linguistic and technical activities are both seen as primitive symbolic forms, used as vectors of a progressive disengagement from the mythical attitude to the world. Cassirer seems therefore to answer the question regarding the relationship of technical activity to language and to mythological thought in a twofold way: first, he identifies the case of technical activity with that of language from the point of view of their mediating power; second, he argues that technical activity is opposed to mythological thought in the same way as language can ultimately become opposed to mythological thought.[4]

Thus, it seems undeniable that the two points of view in *The Philosophy of Symbolic Forms* and in 'Form and Technology' are in opposition to one another, even if language keeps its fundamental role in both of them. Is it only a difference of presentation? The discrepancy seems much deeper than this. The description developed in *The Philosophy of Symbolic Forms* is, in fact, transcendental: language is construed as the basic form within which other symbolic forms can be conceived on an a priori basis. Even if the description starts with language and ends with science, there is no finality which would lead from mythological thought to science as the ultimate symbolic form,[5] and both of them appear more as examples of symbolic analysis than as steps towards an ultimate knowledge conceived in a kind of Hegelian way. The point of view developed in 'Form and Technology' is historical: the mythical interpretation of the world is the basic socio-semiotic situation in the history of humanity, from which a way out became possible thanks to the mediation of language and technical activity. It is the inner relationship between language and technical activity which prompted the mutation leading from a mythical attitude to the world to a scientific one. Therefore, the two viewpoints exposed by Cassirer do not match, not only because finality is not interpreted in the same way in the two works but because the emphasis on language in the first one involves a semiotic state which is immediately given thanks to the presence of signs conveying meanings, contrary to what happens in the second one, where technical activity is not immediately semiotic – that is, is not characterized as using conventional signs the way language does. What seems to characterize technical activity in general is its neutralization of the expressive dimension, due to the fact that it operates unconsciously: even if the final goal of a technical activity can be represented and evaluated by the individual, the technical activity itself is embodied in such a way as to be forgotten entirely. For example, nobody has to think of the potential difference in wires to switch on the light: this is typical of a technical activity in which the process activated to reach a goal is entirely blind. And science in general is precisely a form in which the main goal of technical activity is directed towards *objectivity* conceived as deprived of any expressivity. The point at stake is therefore the following: how is it possible for technical activity, the inner workings of which are deprived of any expressive dimension, to be recognized eventually as an activity in itself and therefore as performing a specific form of expression?

We are therefore faced with two problems: (i) if a symbolic form is symbolic thanks to the semiotic presence of signs only, in what sense can technical activity be considered as a symbolic form in its own right?; and (ii) if technical activity is nonetheless acknowledged as a symbolic form, what is the relationship between language and technical activity? There is, I think, a way to answer both questions which remains faithful to the spirit, if not to the letter, of Cassirer's philosophy.

Activity as the basis of language and tool-making

As for the first question, language is an *activity* just as much as technical activity, as Cassirer pointed out repeatedly, borrowing the Greek concept of 'energeia' from Wilhelm von Humboldt and Karl Bühler in order to characterize its nature. It is therefore because both language and technical activity are activities that they have a common ground, not because language makes use of signs whereas technical activity does not necessarily do so. What does 'activity' mean in this context? An activity is a regular sequence of actions focused on a collective goal which is performed for its own sake. It involves a way of performing specific gestures[6] in which the use of linguistic signs can be included. The collective aspect of an activity should not therefore be interpreted in terms of one-to-one transactions between individuals, since it is not the more or less large number of individuals which makes the activity collective. It becomes collective when it is performed for its own sake, defining by its own process what should be considered as proper for the activity under way. Once an inherent norm is recognized in specific marks and gestures, these marks and gestures acquire a social status: the marks may then be interpreted as signs and the gestures as involving tools. From this moment on, they can be used for their own sake, defining the norm which should be followed by the activity. It is therefore not the fact that tools are external objects extending an activity already achievable by other means like the movement of a limb (arm, hand, finger and so on) that makes them tools; in order to take place, a human activity needs social landmarks that must be recognized as such by those involved in the activity. These landmarks can eventually become tools when the landmarks are involved exclusively to reach the goal of the activity itself; hence the fact that tools are used unconsciously, except when they become dysfunctional. And it seems to me that the same holds true too for the use of linguistic signs, as some psycholinguists have already pointed out.[7]

If language and technical activity are to be associated, as is the case in 'Form and Technology', we must expand the notion of a symbolic form to any kind of shared activity, semiotic or semiotic-to-be. Consequently, we can interpret the nature of linguistic activity as the symbolic form which keeps transforming itself through the interface it has with other activities, like the technical one. It is through this interface that new symbolic forms evolve in their own right, like science. Language remains therefore the fundamental basis from which other activities become susceptible to bearing some expressivity, thus becoming symbolic forms. It is, I believe, this *internal symbolic drift* which is the core of *The Philosophy of Symbolic Forms* as well as of 'Form and Technology', and, more generally, of Cassirer's published philosophy.

As for the second question, which deals with the specific relationship between language and technical activity, it boils down to the matter of how the technical activity becomes a symbolic form. We have to focus on the definition of technical activity given by Cassirer to answer this question.

Three points to reconsider

If technical activity, along with language, plays a mediating role which gradually involves a withdrawal from the mythical attitude to the world, how is this role to be conceived of? It seems to me that at least three points should be re-examined in Cassirer's position if we want to take into account recent advances in the nature of technical activity.

Language and tool-making as criteria for humanity

In 'Form and Technology', Cassirer, along with most of his contemporaries,[8] defends the idea that both language and technical activity are criteria that can be used to characterize humanity as such. This question has received close scrutiny in the last 20 years or so, and, in view of these recent findings, it is not possible to expressly defend Cassirer's point of view today.

Let us take the case of technical activity first. The debate concerning a possible 'animal technical activity' oscillates between two extremes: either the cases of humans and of superior primates are identified by promoters of the animal cause or they are separated by defenders of a human specificity. There are very good arguments in contemporary archaeology and ethology to claim that this second viewpoint is just wrong: technical activity has been tracked back to pre-humans[9] and, to some extent, even to animals. Consequently, the position held at the time of Cassirer is not sustainable in the same terms any longer. But this new state of affairs is an opportunity to clarify the nature of technical activity. Thus, the point under discussion is less the presence or absence of technical activity in animal groups, a fact which is not questionable today, than the way animal technical activity is organized and connected to other activities. The indirect consequence of this last point is that the notion of a symbolic form has to be refined as well.

Let us point out first that in the debates concerning animal technical activity, and more generally concerning animal culture, the extensive use of ill-defined and polysemic concepts such as 'culture', 'tradition' and 'cooperation' makes it almost impossible to validate any serious advance on the issue. But we should try nevertheless to clarify them by going back to some anthropological definition of technical activity, like the one proposed by Marcel Mauss as early as 1936,[10] and compare it to recent findings in ethology, as suggested by anthropologist Frédéric Joulian.[11] The definition of technical activity given by Mauss is based on three features: it is an *activity* which is *embodied* and based on *tradition*. Joulian points out that Mauss could only take into consideration the first feature – that is, the notion of activity – which in his times was studied among superior primates. But today, says Joulian, contemporary primatology is able to track down traditions of behaviours among specific animal groups (such as the famous example of the potatoes being washed by a specific group of Macaque monkeys), as well as embodiments of special ways of performing gestures. Far from resting upon external and rather mechanical criteria to define what a technical activity is, such as the difference between a bodily gesture and an externalized tool, primatology and paleoanthropology are now focusing on what seems to be the collective dimension of activity. But has this collective dimension a *social* aspect in the case of animal groups?

In actual fact, there is a difference in the collective aspects of technical activity in animal and human groups. What seems to be lacking in animal technical activity is the technological *heterogeneity* which is found in human activities:[12] once a specific tradition of embodied activity has emerged within a particular animal group, it spreads over the entire group, just as a chemical reaction would do, and is transmitted as such. Contrary to what happens in the case of animal technology, human technical activity keeps transforming itself while maintaining nonetheless a certain stabilized form. For example, this is the case with the different styles of stone arrowheads, the evolution of which can even help determining a chronology of prehistoric civilizations. On the contrary, animal technical activity seems to immediately reach a standard norm which is not subject to change or in which the change seems to appear randomly. Therefore, it is not the technical activity as such which introduces a difference between humans and animals, but rather the way embodied and transmitted activities are collectively lived, either in a purely symmetrical mimesis as in the animal case or in asymmetrical relationships defining roles among a socially diversified group of individuals as in the human case.[13] Technical activity is therefore not human-specific but the way it is merged in a social network of activities is very likely to be. This has at least one consequence for the notion of a symbolic form: it is not by any activity arbitrarily limited to humans that a symbolic form should be characterized, but by a specific connection between instability and stability in its inner features. It is this connection which can help elucidate the notion of a symbolic drift that I mentioned above.

We must, therefore, add one feature to the definition of technical activity proposed by Mauss: technical activity is an embodied activity based on tradition and depending on a social, and basically invisible, organization in which roles between humans are heterogeneous. We can therefore introduce a difference between the technical activity displayed by animals and humans, on the one hand, and the *technology* only displayed by humans on the other. This way, we can stress the difference between technical activity which is not considered as symbolic yet and technology which belongs to the domain of symbolic forms.

Thus we can say that, even if the position held by Cassirer is now out of date, it does not mean that the debate on the status and the extension of technical activity in the human and animal world is over. The term 'technical activity' can therefore be attributed to a dimension of experience in animal groups but this activity cannot be akin to the technological one developed by humans, not because the notion of a tool would not be used properly in animal groups but because the social dimension in which tools are used seems to be lacking among them. It is therefore possible to draw a continuity between animal and human groups from the point of view of technical activity but only if this caveat is kept in mind: technical activity has not only to do with tools but also with a social experience of the world which, in the present state of knowledge, is thoroughly different in animal and human societies.

Let us briefly mention the case of language now. Language is considered as human-specific in Cassirer's philosophy. This question has stirred a very passionate debate in linguistic circles, as old as the one concerning technical activity and for the same reasons. Though the problem is immensely complex, it would be, however, a mistake to claim that language is spread among other species than humans, even if some species like apes, parrots and many other ones are able to display the ability to communicate. At least two reasons could be given to stress the specificity of the human case in the animal world. First, the fact that many species have evolved modes of communication does not mean that these modes can be assimilated with language. On the contrary, it would be very surprising, from an ecological point of view, to find other species having developed modes of communication that were not fully adapted to their own ecological niche. The human niche is by no means a *terminus ad quem* in the global evolution of species, and other species do not tend to join the human case as if it were the ultimate goal of evolution. Second, I am yet to be convinced that the kind of ability to communicate other species have evolved can be characterized as linguistic: the fact that an ape seems to be able to categorize different kinds of predators,[14] or that a primate can use a word or even a sentence to refer to material objects or feelings like pain or hunger, is still very far from a linguistic performance which has ultimately to do with an intentionally shared social world.[15] As Michael Tomasello has pointed out, a linguistic sign emerges only when it is shared by several individuals in a very complex intentional framework in which every individual in a given situation is able to project his or herself in everybody else's place, in order to view the same object or situation which is referred to but from several points of view at the same time.[16]

The purpose of these remarks is not to suggest that the problem of determining what is linguistic in the animal world is out of date. But what seems to be now more promising is the recent discovery that some linguistic features one would consider as necessary for language to be identified as such are in fact spread over several species in the animal world. Tomasello and his collaborators very clearly showed the existence of cognitive limits between humans and primates both from a psychological and linguistic point of view,[17] but also showed that these limits are not the same when humans and dogs are compared with one another as when humans and other primates are concerned.[18]

What seems therefore to be specifically human is rather the synthesis of features otherwise spread over several species in the animal world, like the act of pointing or the sensitivity to attentional states. It is of course the way this synthesis becomes effective that has to be determined precisely. But whatever the features comprised in this synthesis, we can infer that they are not contained in some unique capacity the humans are miraculously endowed with, but more likely are found within a particular social structure which is unique to them. If this is the case, it is not the linguistic features *per se*, but their synthesis in a single unity, which makes the human language qualitatively different from other modes of communication. This conclusion seems, therefore, to lead in the same direction as the one which was reached when the nature of technical activity was discussed.

Technical activity as organ-projection

In 'Form and Technology', Cassirer, reinterpreting in an entirely different way speculative ideas first developed by Ernst Kapp, suggests that the artificial extension of a bodily organ provided by a tool implies a progressive self-recognition of its function as a mediation in various contexts. Cassirer's idea is that using a tool is also transforming its usage, committing the individual to the ever-increasing power of mediation by a progressive detachment from the immediate environment in which it was first conceived. It is this last process that Marx, as Cassirer remarks, rightly called 'emancipation of the organic barrier'.[19] But I wonder if Cassirer – along with Marx before him – does not actually give too much credit to Kapp in acknowledging the existence of a kind of primitive 'natural state' of the body, in which individuals would be bound to their organic selves only and from which humanity parted thanks to the mediation of technical and linguistic activities. For what would this 'primitive state' of the body be if it was deprived of language and tool-making? As I said before, the long process through which proto-humans finally became the speaking and tool-making species we know did not have to start from scratch, triggered by a cognitive capacity suddenly becoming active and generating all the cultural changes in the subsequent history of humanity. Without projecting on a distant and imaginary past a 'state of nature' of the body which reminds more of Rousseau than of current anthropology, we must rather acknowledge that the notion of a bodily organ which is in question has very little to do with the naturalized image biology has provided us with. In fact, it has more to do with mythical and linguistic traits that are always present in human cultures. Let me give an example.

Clarisse Herrenschmidt showed, in a very Cassirerian way, so to speak, that the invention of writing as a tool used for the transcription of oral languages in Mesopotamia took place when the organ of speech was externalized under the appearance of a speaking mouth made of clay on which signs were engraved.[20] Much later, in ancient Greece, the invention of coined money in the sixth century BC followed a similar process: it is through the representation of an eye able to see and evaluate that coins first appeared.[21] But her discovery is not referred to a naturalized organ-projection as is described by Kapp, but to the myths and states of knowledge current in Mesopotamian and Greek cultures at the time of these technical inventions. Thanks to a patient reading of the Mesopotamian myth describing the origin of writing, as well as passages from Herodotus referring to vision and, in a concealed way, to money, she was able to show that it was the mythical organs as they were imagined in archaic Mesopotamian and Greek cultures that laid at the heart of the technical role bestowed to the organs.

In the particular case of the emergence of coined money in ancient Greece, which will be, for obvious reasons, of more interest later on in the discussion, two cultural facts seem to have played a part in the connection between the eye and the valuation of goods. First, the eye was considered at the heart of the process of vision: nobody would guess today that the state of knowledge at that time was that the human eye was endowed with the capacity of throwing material rays enabling vision and that the notion of light as it has been commonly used in physics since the seventeenth century had absolutely no place in this process, except for expressing colours.[22] Second, a semantic network in ancient Greek connects (i) the goddess Artemis, (ii) the eye, the coin and the moon, all of them being compared to shining disks, and (iii) the process of measuring and evaluating. Let me briefly describe this network. The rounded shapes of the moon, the eye and the coin have obvious morphological similarities. But they also glitter from the inside since they do not receive their brightness from an external source, such as light for us today. As for the moon, its etymology in ancient Greek refers to measuring and evaluating since 'moon month' is 'that which measures', as Cassirer, among others, has pointed out.[23] Even if this etymology is still philologically controversial today, archaeology can be of some help on this particular matter: the moon is an attribute of Artemis since a very remote antiquity and it is in the temple of Ephesus devoted to Artemis[24] that the earliest coined money ever found was discovered during the 1904–6 excavations directed by archaeologist David George Hogarth. There seems to be, therefore, a connection between the cult of Artemis as goddess of the moon and the use of money. The act of measuring and evaluating which is necessary for the exchange of goods could then be associated with vision through the morphological motif of the shining disk, joining together the moon, the eye and the coin. There is therefore enough evidence to show that, contrary to Kapp's point of view, it is not the artificial extension of a bodily organ only that can explain the progressive self-recognition of its function as a technical mediation in different contexts, but a specific cultural and social background – technological, linguistic and cognitive. This example amply shows that there is just no 'natural state' prior to a hypothetical moment where an 'organ-projection capacity' the existence of which we have to bet on would activate such mediations as language and tools. In fact, there is no way in which a state of human activities could be conceived of as deprived of mediations.

This being said, the idea of organ-projection is not to be entirely dismissed, for it is true, as Cassirer points out in 'Form and Technology', that it is related to the idea of self-knowledge, even if today's advances in psychology would interpret this relationship in a very different way.

Technical activity as an anticipation of self-knowledge

In the case of technical activity, Cassirer insists upon the fact that the individuals encounter from outside something which is, in fact, unconsciously produced by them and in which they can, later on, recognize the mark of their own self. We already noticed that this was the way in which Cassirer conceived of the progressive mutation of technical activity from a globally unconscious activity to a consciously oriented one, that we called 'technology'. The reason why technical activity is interpreted by Cassirer as an anticipation of self-knowledge is therefore that it is the only way technical activity can become progressively integrated in the realm of symbolic forms. But since I have just shown that this progressive integration cannot be the result of an organ-projection capacity only, is it still possible to construe technical activity as an anticipation of self-knowledge? Can we conceive technical activity as an activity which progressively becomes consciously oriented, just like any another symbolic form? And if this is so, does this process follow the same direction as language?

From unconscious to partially conscious activity

In the case of technical activity, the steps leading from an unconscious activity to a subjectively conscious one are not as straight as Cassirer seems to assume. There are at least two reasons why technical activity keeps being a challenge for a subjectively conscious reflection of its own process: (i) in technical activity, the projection of the self made possible by tool usage implies an embodiment which has necessarily an unconscious phase; (ii) technology cannot be mastered by a single individual and is the result of a global social structure in which the individual plays only a small part.

First, the incidence of technical usage of self-knowledge seems doubtful at an individual level: for a technical device to be efficient, it must be unconsciously incorporated and must therefore become a part of the individual's body. Let me take an example borrowed from psychologist Charles Lenay: if I try to park a car and if I hit the pavement, I perceive that 'I' hit the pavement, not the wheels of the car which became part of myself during the process of driving. Drawing from this example, Charles Lenay makes the following remark: 'when a tool is used to perceive, it is not itself perceived. The tool does not participate in the perceptive activity as a perceived form but as it transforms the conditions of action and therefore, all the perceptive field accessible.'[25] Therefore, it is true that the use of a technical device unconsciously expands the self by way of what can be called an 'organ-projection' and it is true that the perception of the self is being modified in the process. But it is only when the device becomes dysfunctional – that is, when the expanded self stops being projected in the device – that the individual becomes aware of this expansion which has now disappeared. It is, therefore, in a very peculiar situation only that the expansion of the self becomes conscious – that is, only when it is *remembered* and not presently perceived through the technical device. We must, therefore, come to the conclusion that there will always be an unconscious moment in technical usage and that the relationship between unconscious and conscious moments in the case of technical usage does not necessarily lead to an additional self-knowledge.

The second reason why technology is not directly connected to self-knowledge has to do with its social aspects: as far as the process of creating technological devices is concerned, it is by no means a rule that it is mastered by a single individual. Although the example of Leonardo da Vinci mentioned by Cassirer in 'Form and Technology'[26] seems to indicate the opposite, since Leonardo managed to use his creative genius in technology as much as in art, even he was not able to master the whole process of research and development that is needed to complete a technological device: most of his engineering projects remained unfinished at various stages of development because they lacked the relevant social structure. Moreover, the example of Leonardo shows that technological creation is interpreted by Cassirer through an Aristotelian perspective, which induces some kind of misunderstanding regarding the difference between individual craftsmanship and socially based technology. But a tool is not a technology, precisely because its conception and usage can be mastered at an individual level, contrary to technology. One can even claim, as Alfred Sohn-Rethel did, that technology rests precisely upon a social division of labour between those who collectively develop and apply science, on the one hand, and those who do not, on the other.[27] Still, self-knowledge entertains some link with technology interpreted as a social activity.

Publicly shared activities as symbolic forms

Self-knowledge does not depend on the private capacity of organ-projection which would find an expression in technology. Self-knowledge depends on publicly shared forms through which individuals find ways to express themselves. This interpretation of self-knowledge supposes that individuals do not naturally possess the semiotic means of expressing themselves but find these means in socially instituted forms, such as technologies and languages, which were present before the individuals and which they inherit and transform. Therefore, self-knowledge depends on socially warranted activities materialized in specific symbolic forms, one of them being technology. Interpreting Cassirer's notion of a symbolic form this way, self-knowledge is not the immediate response one can naturally expect from technology usage: self-knowledge in general is a socially mediated effect of several symbolic forms on the individuals. The case of technology, both from the point of view of usage and of creation, is particularly clear on this point since its whole process cannot, as a social activity, be mastered from an individual point of view. A 'symbolic' activity will therefore be broadly defined as an activity which makes possible the organization of social behaviours and anticipates their course via publicly shared forms, symbols and values at the same time.

These three points being clarified, I would like to study the relationship between technical and linguistic activities in a particular case: the analogy between money and language.

The analogy between money and language[28]

First, let me justify the use of this analogy by quoting the third volume of *The Philosophy of Symbolic Forms*, where Cassirer extensively recalls the Kantian phrase according to which modern science is able to 'spell out phenomena so that we may be able to read them as experience'.[29] If we take the image literally, modern science started when nature became readable. The comparison between the letters of the alphabet and the atoms of the universe is a recurring theme since the beginning of philosophy in ancient Greece, as Eric A. Havelock has pointed out.[30] Though this is not explicitly stated by Cassirer, it is the technology of writing and reading which contributed to the severing of language from myth and its redirection towards what would become science. If we agree in saying, as Cassirer maintains in 'Form and Technology', that the tool announces the twilight of the magical and mythical world,[31] then the comparison between technical activity and language should focus on one of its most powerful means, the graphic technology of writing. Writing construed as a graphic technology is used for many purposes, even if what comes immediately to mind is the translation of oral speech. But another usage is of equal importance: the possibility of describing and extensively using the notion of numbers. In the third volume of *The Philosophy of Symbolic Forms*, just mentioned, Cassirer states also, quoting pre-Socratic philosopher Philolaus, that it is in the mastering of the realm of numbers *per se* that science originates.

Therefore, I suggest that, rather than studying the relationship between the technology of writing and language interpreted as oral speech only, the comparison should also include the relationship between the technology of writing and the notion of numbers, considered as the main semiotic medium of science. In this way, language can be studied from a technical angle and in its full range, from its mythical to its scientific usages. That is why the comparison between language and technical activity that I would like to defend in the following pages as a case study concerns the notions of *money* considered as a technological vector of arithmetic, on the one hand, and *language* as the most fundamental symbolic form on the other. This way, I hope to provide an example of what makes possible the 'symbolic drift' I referred to above.

Three common features

Contrary to a classical viewpoint, which would consider that either the monetary value or the linguistic meaning depends on the intrinsic nature of the entity they refer to, the notion of a symbolic form helps clarify the fact that it is through the transactions themselves that the very notion of value or meaning can emerge. This is, I think, what Cassirer calls a 'mediation': an activity produces its own material medium but it is through this medium that the activity keeps being elaborated. From this point of view, money and language are not different: as material mediations, any monetary value or linguistic meaning is indistinctively a sign of a shared interaction and a tool for its investigation. This has several important consequences.

General equivalent

To make myself clear, I will start with a very simple model of exchange first developed by anthropologist Alain Testart.[32] Let us suppose that three persons A, B and C exchange different goods with one another in one-to-one transactions: A exchanges with B, B with C and C with A. In most cases, after a certain period of time, no one will have exchanged exactly the same quantity of goods and individual A can be in debt to individual B, as well as B to C and C to A. If they finally decide to use a certain kind of monetary token in order to get rid of the debt each of them has towards another one, they will therefore replace a contextual debt related to a specific person and a specific transaction with a decontextualized situation in which neither the persons nor the transactions matter any longer. Money transforms personal relationships into arithmetic quantities as long as the tokens of these quantities are recognized and trusted as such by all the parties. As soon as a monetary token becomes decontextualized, the value given to the token depends on multiple transactions which all depend on the expectations of the different protagonists. Money becomes, therefore, a general equivalent as it gives access to any kind of goods, whether present or not, for it anticipates the attribution of any value to any good on a predetermined scale.

This is also exemplified in the history of coined money in ancient Greece: used first in the *Artemision* as a propitiating token given by women before childbirth, the use of money would expand in other directions of activity, reformatting the notion of exchange itself and imposing its own standard of monetary exchange.[33]

This is also true of language: the attribution of a particular linguistic meaning to a sign depends on the collective use of this meaning, which is not only related to the context in which it is presently in use but also to a collective usage that plays the same part as a market governed by supply and demand. And, in both cases, any monetary token or linguistic sign, once it is recognized and trusted as such, anticipates the value or meaning of any other token or sign, whether it is present or not. But, of course, language is a much more complex form than money, for the latter lacks the very intricate compositionality that one finds in languages and which cannot be accounted for on a simple arithmetical scale.

General equivalents are only tentatively universal: it is through the diversity of languages that something like a meaning can be conceived of, as it is only through the operation of monetary exchange that distinct currencies keep a differential value and remain valuable. Consequently, what appears to be an instability of value or meaning is not a shortcoming but the very condition of possibility of their existence: it is necessary that interactions modify the open series of their occurrences if a value or a meaning is to remain alive.

Self-evaluation

Since a monetary value or a linguistic meaning does not depend on a predetermined nature derived from the entity (thing or good) it is attributed to,[34] it must be within the transaction in which it participates that it takes shape. As a result, an activity defines its own criterion of evaluation because it modifies its own shape by modifying the internal medium it has itself produced. To this extent, any activity involving money or language as material mediations is at the same time, as Cassirer said in 'Form and Technology', an activity focused on money and language themselves, interpreted as a sign of a social interaction and as a technology for its investigation. This is made clear by the example I have already used: the progressive extension of monetary use in ancient Greece changed the very nature of what was considered valuable because what was exchanged was, henceforth, a measurable good, sold or bought by protagonists who then became private individuals taking part in a market (*agora*), itself warranted by one or several third parties (market supervisors appointed by the city called *agoranomoï*,[35] state mint and customs).[36] Starting from a ritual and religious context, coined money became the archetype of an exchange of goods. Consequently, what is considered valuable depends as much on symbolic values as on 'useful' ones because the very notion of utility is not defined once and for all but emerges from the transactions it makes possible. Therefore, any monetary or linguistic theory which presupposes that a predetermined and objective criterion can be assigned to an object or a sign – for example, utility in economics or logical reference in linguistics – is bound to be criticized on the same ground as Kapp was criticized by Cassirer, for there is no objective nature from which values and meanings can be derived prior to the effective activity itself.

Practical and mythical aspects

The fact that what is valuable or has a meaning cannot be determined in advance outside the effective transactions it is engaged in, casts light on what Cassirer means by the 'mythical' aspects of tool-making and language: practical and mythical aspects of value and meaning are intertwined and cannot be severed from one another. Let me start with money first, in which the fictional and practical dimensions are clearly intermingled.

Money is fictional because its purchasing power entirely relies on trust and mutual anticipations between the individuals who accept its role and are therefore engaged in a specific kind of transaction which was non-existent before it was set up. Moreover, when the historical origins of money are examined, one finds that it was not the utilitarian perspective which was the key factor that triggered its emergence. Money appeared first in a mythical and ritual context in order to respond to social commitments that had no mercantile basis: marriage, mourning, vengeance, favouring the gods. On the other hand, money cannot be considered as fictional only because, as a general equivalent, it completely revolutionized the structure of exchange itself by transforming the very idea of what an exchangeable good was. From this point of view, it is a constraint placed upon every other individual exchange once it has been set up as a standard form. There is, therefore, a strong continuity between social obligation, mythical participation in the world and economic exchange.

The same is true about language: language is fictional in the sense that the meanings attributed to signs entirely depend upon a mutual agreement which is arbitrary; but language cannot be considered as fictional only, because it also participates in the production of things, actions and social roles and is not merely a recording of what would go on otherwise. Language is more an ongoing drama in itself than a way of reporting what happens outside itself. This drama cannot be founded only on the search for pre-existing meanings because meanings, the search for which is described in language, are themselves built and concretely accessed through language activity.

Following Cassirer, who showed many times that, in order to scientifically describe an object, one has to abstract its functional role, I will now describe the four main functions that money usage and linguistic activity have in common.

A chart of four common functions

Four traditional functions of language can be analysed by analogy to those classically assigned to money: evaluation, payment, circulation and saving. They all can be described in a chart which shows tight correspondences between the two kinds of activity:

	Money	Language
Evaluation	– Anticipation of supply and demand – Differential valuation of goods – Money as a means of evaluation is part of evaluation itself	– Anticipation of future usage – Differential meanings determined through predication – Language as a means of evaluation is itself being re-evaluated
Payment	– Medium of decontextualized values that can be recontextualized – Diversity and competition between currencies according to the type of transaction – Status: ex creditor/debtor – Obligations: contracts, debts – Roles in transactions	– Medium of decontextualized meanings that can be recontextualized – Diversity and competition between types of discourse according to context – Status of addressees – Obligations: stylistic codes and genres – Actantial roles
Circulation	– Conversion of things into commodities and goods – Diversification of means of transaction (money, cheques ...) – Perception of flows (money, commodity) by agents	– Sharing of common experience structured in 'objects', 'actions', 'qualities' through naming and predication – Diversification of meaning (polysemy) and development of vocabulary – Perception of thematic genres by speakers
Saving	– Hoarding – Authorized institutions (states, banks): warranties and norms – Standardization	– Vocabulary – Idiomatic phrases, proverbs – Authorized speakers – Canonical forms of discourse

This chart shows how precise the correspondence between money and language can be. From a philosophical point of view, it shows also in what sense language plays a fundamental role in the gradual differentiation of symbolic forms, since it remains the basis from which other forms emerge.

Conclusion

The comparison between money and language developed here is just one point of view among many others from which Cassirer's conception of technical activity can be evaluated. I hope this example has contributed to show the following: (i) that the general framework set up by Cassirer allows the progressive building-up of technical activity as a symbolic form; (ii) that, nevertheless, contrary to what Cassirer tends to do in 'Form and Technology', it is not possible to restrict the nature of technical activity to the notion of tool usage, for it also involves the gradual constitution of technology based on semiotic interactions; (iii) that, more deeply, Cassirer tends to minimize the expressive dimension of technical activity, although this dimension is perceptible in the notions of style and norm (one reason which would explain this minimization is that technical activity contributes to the objectivity of science in building up experiments deprived of any expressive dimension, science being acknowledged by Cassirer as one of the most fundamental symbolic directions taken by humans); (iv) that it is possible, however, to show that the interactions between technical and linguistic activities are deeper than one might think first, as the comparison between money and language has, it is hoped, revealed.

Notes

1. This article was first conceived as a contribution to the workshop organized by Aud Sissel Hoel and Ingvild Folkvord, which took place at the Nordeuropa-Institut at Humboldt University in Berlin in February 2008. It is also part of a research project funded by the French Agence Nationale de la Recherche under the title 'Semiotic Perception and Sociality of Meaning' (ANR-06-BLAN-0281). The following pages benefited from the collective work carried out in Berlin (with Ingvild Folkvord, Aud Sissel Hoel, John Michael Krois, Tord Larsen, Steve Lofts and Frederik Stjernfelt) and in Paris (with Pierre Cadiot, Clarisse Herrenschmidt, Patrice Maniglier, Victor Rosenthal, Chris Sinha and Yves-Marie Visetti). I would like to thank them for their inspiring comments and cooperation, as well as Pierre and Darla Gervais for their help in improving my English.
2. In 1925, Cassirer published not only the second volume of *The Philosophy of Symbolic Forms*, focusing on mythical thought, but also the article 'Sprache und Mythos: ein Beitrag zum problem der Götternamen', which I consider Cassirer's best analysis of the relationship between language and myth.
3. That is, only one year after the last volume of *The Philosophy of Symbolic Forms*. Cassirer's personal development on the question of technical activity seems to have been influenced by his 1929 meeting with Martin Heidegger in Davos.
4. From this sketchy outline, one could easily be led to conclude that the symbolic evolution of humankind consists in the progressive separation from

mythological thought and its replacement by science. But this is not the case: if it were, language would stop occupying the transcendental core in the system of symbolic forms and would eventually be replaced by science, the development of which is based on a divorce with natural language, as Cassirer often pointed out. But this replacement never happens, for linguistic activity is the unique driving force able to mingle into all human activities in general.

5. We know that Cassirer prepared a sequel which was published in 1995 under the title *Zur Metaphysik der symbolischen Formen*, vol. 1 of *Nachgelessene Manuskripte und Texte*, ed. John Michael Krois (Hamburg: Meiner).

6. A well-documented example is the social difference between the right and the left hand. See Roger Hertz, 'La prééminence de la main droite: Étude sur la polarité religieuse', *Revue philosophique*, LXVIII (1909), 553–80.

7. See Michael Tomasello, *The Cultural Origins of Human Cognition* (Harvard: Harvard University Press, 1999).

8. A notable exception should be made for the Gestalt psychologist Wolfgang Köhler who published in 1921 an important book on ape intelligence, *Intelligenzprüfungen an Menschenaffen* (Berlin: Springer).

9. See, for, example, Sally McBrearty and Alison S. Brooks, 'The Revolution that Wasn't: A New Interpretation of the Origin of Modern Behavior', *Journal of Human Evolution*, 39 (2000), 453–563. I will not consider the case of pre-humans here, for it would take too long. An analysis of this case can be found in Jean Lassègue, 'Introduction', in *Emergence de la parenté* (Paris: Editions Rue d'Ulm, 2007).

10. Marcel Mauss, 'Les techniques du corps', in *Sociologie et anthropologie* (Paris: Presses Universitaires de France, 1950), 365–86.

11. Frédéric Joulian 'Techniques du corps et traditions chimpanzières', *Terrain*, 34 (2000).

12. This idea was defended by Yves-Marie Visetti in the Parisian workshop already mentioned, 'Perception sémiotique et socialité du sens'.

13. It seems that the notion of group coalition which has been studied extensively among primates does not fulfil the social requirement I just mentioned: it is a way of solving a crisis between two individuals by forming coalitions on each side, however strong the two individuals are in the beginning. It is not properly social in the sense I gave to the term, for there is no object of transaction which would be pursued for its own sake.

14. This very famous example is still a matter of debate.

15. Michael Tomasello, 'Why Apes Don't Point ?' (plenary talk at the 5th Evolang Conference, Leipzig, 31 March–3 April 2004).

16. 'To learn to use a communicative symbol in a conventionally appropriate manner, the child must engage in what I have called role-reversal imitation. That is, the child must learn to use a symbol toward the adult in the same way the adult used it toward her. ... The child's role and the adult's role in the joint attentional scene are both understood from an "external" point of view, and so they may be interchanged freely when the need arises. ... The result of this process of role-reversal imitation is a linguistic symbol: a communicative device understood intersubjectively from both sides of the interaction.' Michael Tomasello, *The Cultural Origins of Human Cognition* (Harvard: Harvard University Press, 1999), 105f.

17. Michael Tomasello and Hannes Rakoczy, 'What makes Human Cognition Unique? From Individual to Shared Collective Intentionality', *Mind and Language*, 18 (2003), 121–47.
18. Joseph Call, Juliane Braüer, Juliane Kaminski and Michael Tomasello, 'Domestic Dogs (Canis familiaris) Are Sensitive to the Attentional State of Humans', *Journal of Comparative Psychology*, 117:5 (2005), 257–63.
19. Ernst Cassirer, 'Form and Technology', in this volume, 38.
20. Clarisse Herrenschmidt, 'Ecriture, monnaie, réseaux: invention des Anciens, inventions des Modernes', *Le Débat*, 106 (Paris: Gallimard, 1999), 37–65.
21. Clarisse Herrenschmidt, *Les trois écritures: Langue, nombre, code* (Paris: Gallimard, 2007), 232.
22. Gérard Simon, *Le regard, l'être et l'apparence dans l'Optique de l'Antiquité* (Paris: Seuil, 1988), 88.
23. Ernst Cassirer, *Philosophie der symbolischen Formen*, vol. I: *Phänomenologie der Erkenntnis* (Darmstadt: Wissenschaftliche Buchgesellschaft, 1929), 257.
24. The Artemision in Ephesus was considered one of the Seven Wonders of the ancient world.
25. Charles Lenay, *Ignorance et suppléance: la question de l'espace* (Habilitation à Diriger des Recherches, Université de Technologie de Compiègne, 2005): 113.
26. Cassirer, 'Form and Technology', 43.
27. He uses the notion of a symbolic form in a rather idiosyncratic way, which is somewhat different from Cassirer's. See Alfred Sohl-Rethel, *Geistige und Körperliche Arbeit: Zur Theorie der gesellschaftlichen Synthesis* (Frankfurt am Main: Suhrkamp, 1970); English translation by Martin Sohn-Rethel (Atlantic Highlands, NJ: Humanities Press, 1978), 122: 'The capitalist control over the labour process of production can only operate to the degree to which the postulate of automatism functions. The stages in the development of capitalism can be seen as so many steps in the pursuit of that postulate, and it is from this angle that we can understand the historical necessity of modern science as well as the peculiarity of its logical and methodological formation. As pointed out earlier in this study, the mathematical and experimental method of science established by Galileo secured the possibility of a knowledge of nature from sources other than manual labour. This is the cardinal characteristic of modern science. With a technology dependent on the knowledge of the workers the capitalist mode of production would be an impossibility.'
28. This part of the article directly derives from Jean Lassègue, Victor Rosenthal and Yves-Marie Visetti, 'Économie symbolique et phylogenèse du langage', *L'Homme*, 192 (2009), 67–100.
29. Immanuel Kant, *Prolegomena zu einer jeden künftigen Metaphysik, die als Wissenschaft wird auftreten können*, § 30, AK 4. 312.
30. For example, Eric A. Havelock, *The Literate Revolution in Greece and its Cultural Consequences* (Princeton, NJ: Princeton University Press, 1982), 80–1.
31. Cassirer, 'Form and Technology', 25.
32. Alain Testart, *Aux origines de la monnaie* (Paris: Éditions Errance, 2001).
33. Clarisse Herrenschmidt, *Les trois écritures* (Paris: Gallimard, 2007), 229f.
34. 'Because wealth is what the others consider as wealth, what matters is not to find a particular thing defined in terms of natural properties, but to imitate

others in order to discover in which direction the collective consensus heads to. Money does not originate from a Contract or from the State but from the mimetic and spontaneous convergence of individuals looking for protection [and for the true definition and value of their desires].' Michel Aglietta and André Orlean, *La monnaie entre violence et confiance* (Paris: Odile Jacob, 2002), 77.

35. The 'nomos' here being a territorial division of the size of a district and not the 'nomos' meaning 'law', which is spelled differently.
36. Raymond Descat, 'Le marché dans l'économie de la Grèce antique', *Revue de synthèse*, 127 (2006), 253–72.

7
The Power of Voice

Ernst Cassirer and Bertolt Brecht on Technology, Expressivity and Democracy

Ingvild Folkvord

An archive recording gives me access to the reading of a German poem. The recording is from 1953. The sound quality is rather poor, yet the words are clear. They are spoken by a male voice whose intonation indicates a connection to southern Germany. I am familiar with the wording from previous readings, but find myself surprised by the intonation: it starts out almost monotonously, then it becomes more vivid, and certain phrases are even spoken with an almost pastoral diction. The recording is of the German author Bertolt Brecht rendering his own poem 'An die Nachgeborenen' ('To Posterity'), which he wrote between 1934 and 1938. The poem addresses future generations, asking them to be forbearing in their memory of a colloquial 'we', who 'wished to lay the foundations of kindness', yet 'could not ourselves be kind'.[1] Towards the end of the recording, in the part which appears as the third unit of the written text, the reader's voice changes significantly for the word 'Ihr' – 'you' – which is read out more loudly and with a stronger emphasis. It is as if, through this emphasis, Brecht wants to contribute to the transmission of his message across the time separating him from his future listeners and readers who are explicitly addressed in this now-canonized German poem.

Bertolt Brecht was consciously aware of the difference it makes to our reception, whether we read a text or hear it rendered by a human voice. As early as the 1920s, he reflected critically on the social effects of radio mediation and, later, his experimentation with the new genre of the radio play coincided with Ernst Cassirer's work on the essay 'Form and Technology'. In Cassirer and Brecht's cultural environment during this particular time, in the politically unstable Weimar Republic of the 1920s and the early 30s, the negotiations on the new mass media were closely intertwined with the political tensions of the period and its different versions of 'faith and mistrust in democratic procedures and in the political

161

maturity of the citizens'.[2] Whereas Brecht was strongly concerned about the potential invasion and seduction of the listener by the new state-controlled mass media, Cassirer's more principled approach to technology and meaning sought to develop a theoretical framework which accounted for modern technology as a potentially formative tool, integrating human expressivity as a level of meaning. A juxtaposition of these two thinkers may seem to amount to a comparison of incompatibles. And indeed, the liberal philosopher and the Marxist author address different audiences; they work in different genres with different aims and ambitions. Yet, they meet in a dynamic understanding of cultural tools and a concern about the development of democracy in a politically polarized cultural environment. Furthermore, there is a strong connection and continuity between Brecht's avant-garde approaches from the first third of the twentieth-century, and later attempts to conceptualize the auditive field in critical theory and poststructuralism. This is the particular context in which my comparison of Brecht and Cassirer seeks to shed light on the resources contained in Cassirer's approach to meaning, and the current relevance of his approach for an attempt to frame and understand voice phenomena and listening reception.

In the first part of this chapter, I address this issue by drawing critical attention to the scepticism which has determined influential systematic approaches to voice phenomena and radio culture since the end of World War II. In the second part, I juxtapose Brecht's take on radio and Cassirer's understanding of technology and meaning. Initially, I compare Cassirer's rather principled approach in 'Form and Technology' – in which he seeks to integrate technology as one of several symbolic forms – to Brecht's critical engagement with the new mass medium of radio in his immediate cultural environment. My comparison of their approaches to human expressivity then concentrates more specifically on their respective attempts to come to terms with the effects of what Marshall McLuhan later referred to as the 'intimate', or the 'person to person'[3] aspect, involved in radio mediation. The final part of the chapter centres on the ambiguous notion of the open subject: on the listening receptivity regarded as a resource, but at the same time as a source of highly problematic exposure to external forces and their manipulative power.

Hardened scepticism

Whereas radio practitioners, audio-book publishers, and other non-academics have emphasized the connection between listening reception and creative imagination, systematic approaches in the fields of cultural studies and aesthetics have been far more ambivalent and reluctant to recognize the particular expressivity of voice phenomena as part of our cultural production of meaning. The intimate connection between medium and listener, and radio's seductive force as a result of this intimacy, has been a major concern in post-World War II approaches to the radio: 'The radio affects most people intimately, person to person,' wrote Marshall McLuhan in 1964, emphasizing the allegedly problematic mythical dimension of these dynamics. McLuhan likens the radio to a 'tribal drum'.[4] He explicitly relates this comparison to one particular chapter in the history of the radio medium, namely its role in the National Socialist propaganda apparatus during the Third Reich. This is a crucial point of reference; McLuhan alleges that, as a medium, radio has the capacity to evoke 'archaic tribal ghosts of the most vigorous brand'.[5]

In a more recent approach, Wolfgang Welsch (1997) asks whether our contemporary culture is fast becoming an auditive one. Such a development would 'intensify our awareness of other people and nature', he comments.[6] Welsch, then, conceives of listening as a mode of reception which involves 'openness to the event'.[7] Although he considers this openness a cultural resource, he sees it simultaneously as highly problematic: 'Tone penetrates, without distance', Welsch claims, and this is why 'we are especially in need of protection acoustically'.[8] When such protective gestures appear to be of particular interest to my investigation, it is because they frequently indicate systematic shortcomings, a limited capacity to deal with expressive sound phenomena as meaningful cultural utterances in their own right. Welsch's analysis remains rather vague. It is unclear how 'we' ought to be protected, by whom and from what. Yet, his approach to the auditive field exemplifies a characteristic ambivalence: the openness traditionally associated with listening reception is given positive connotations, but the same openness is also identified as the very factor which allows violent manipulation by an external force.

Similar tendencies can be traced in the field of aesthetics – for example, in Mieke Bal's (2002) analysis of voice mediation in James Coleman's installation *Photograph*.[9] Although Bal elaborates extensively and interestingly on the voices in Coleman's installation, at the systematic level of her analysis she nonetheless adheres to a conceptualization of the listening subject as passive recipient. Or, more precisely, he is passive unless he is empowered – brought into an active role – as a subject participating in the work of art. According to Bal, such participation can be achieved through formal techniques which effectively block the seductive proximity and the processes of identification produced by the mediation of personal voice. In her reflections on Coleman's work, the 'heterochrony' between words and images – the fact that the schoolgirls in the images do not speak themselves, but are accompanied by a 'theatrical and historical'[10] voice – emerges as a decisive formal element. This helps create a 'suspension of figurativity',[11] Bal argues. Within her framework, this suspension appears as aesthetically and politically potent, since it serves as a precondition for the ability of this particular work of art to contribute to an effective 'anti-individualistic'[12] exploration of subjectivity. The analysis thus favours artistic strategies of depersonalization. These play a vital role in contemporary aesthetics, as they did in the avant-garde radio discourse of the Weimar Republic: 'The radio speaker ... has no other function than the type,' Rudolf Arnheim insisted in his radio reflections from the early 1930s,[13] as if rigid formalism could eliminate the expressivity of voice phenomena and the physiognomic dimension involved in our listening reception.

One significant connector between such avant-garde approaches of the first third of the twentieth century and these later attempts to conceptualize sound issues is the idea of separation and distance as the precondition for productive understanding, as reflected in Brecht's concept of the 'Verfremdungs-Effekt' – also known as alienation, estrangement, or the V-effect. The Brechtian V-effect is related to strategies of depersonalization, and is a well-known part of Brecht's epic theatre, which he developed from the 1920s onwards. Brecht's idea was to create modes of presentation which would limit audience empathy; to reduce the spectator's potential identification with the characters of a play through a conscious distancing. In order to prevent the audience from sinking into the illusion of the play, Brecht suggested that specific techniques be applied. These included the backdrop projection of pictures or texts; use of placards; episodic presentation of events; and particular modes of acting which emphasized the character as a role, thus highlighting the fictional status of the theatrical play. Together, such techniques aimed to produce a distance which would enable the audience to reflect critically upon the play. Although primarily associated with Brecht's theatrical work, the idea of the V-effect also has strong roots in his work on the radio medium, as we shall see. And, like the approaches taken by McLuhan, Welsch and Bal, Brecht's critical reflections on listening reception are inseparably tied to reflections on cultural agency and democracy.

Yet, there are significant differences between Brecht's critical practice during the late Weimar Republic, and post-World War II academic approaches to the auditive field, such as the ones referred to above. First, these more recent thinkers do not share Brecht's ideologically determined notion of how art should contribute to institutional and social change. Second, their approach to language and media technologies is different: throughout the twentieth century, one can trace a gradual radicalization and hardening of the scepticism inherent in Brecht's avant-garde reflections. Influential media theorists from the second part of the twentieth century, such as McLuhan and Friedrich Kittler, typically subscribed to a media hermeneutics of scepticism, as opposed to a more formative cultural hermeneutics emphasizing human agency and active reception. Different versions of anti-essentialism and anti-intentionalism have together contributed to a paradigm in which the listening subject is frequently conceptualized as a target, a potential victim, or even as the site of technological warfare. In an approach conflating the history of media technology precisely with the history of warfare, Kittler claims that, rather than contribute to the development of human culture, the new technological possibilities to store optical and acoustical data have produced a development in which 'human memory capacity is bound to dwindle'.[14]

Today, as the paradigm to which such radicalized media scepticism belongs is about to be questioned in highly productive ways, the French philosopher Jacques Rancière points to one of the problematic aspects shared by many of these approaches, namely their definition of the listener (and the observer) as a passive receiver – unless he is activated by certain formal strategies, or, as a theoretician or artist, inhabits a position from which it is seemingly possible to act and intervene in a variety of ways. Rancière defines such a critical practice as a 'police distribution of the sensible',[15] questioning it as a highly problematic regulation of the field. The active roles are reserved for art and theory, whereas the ordinary listener, reader, or observer is defined as a site of passive consumption.

On this issue, Rancière's agenda resonates with Cassirer's cultural theory, in which agency is distributed far more generously due to his recognition of human expressivity as a significant and indispensable level of meaning. As a consequence, frequently denigrated notions, such as figurativity, personification and identification – which belong to the realm to which both Brecht's strategies of 'Verfremdung' and later versions of critical theory and poststructuralism have sought to establish a critical distance – can be framed as part of the human production of knowledge. Or, as put by Cassirer in his essay 'Form and Technology', they can be conceived of as part of the process through which the human 'progressively builds up his world, his horizon of "objects"'.[16]

Reading Ernst Cassirer's essay from the vantage point of the present, and coming from a field of cultural studies still defined by a media hermeneutics of radical scepticism, his work provides a very different approach to technology. This is especially apparent in its repeated insistence on the principles of active reception and formative development. Drawing heavily on classical references, and including the tradition of Kant, Goethe, Schiller and Humboldt, Cassirer asserts, time and again, the individual's capacity to contribute productively to the building of his world: 'The "form" of the world, whether in thought or action, whether in language or in effective activity, is not simply received and accepted by the human being; rather, it must be "built" by him,'[17] Cassirer claims. Hence, he juxtaposes linguistic and technical activities as processes which are both equally important in relation to what he sees as the gradual formation and articulation of the human world. Although one might rightly object that Cassirer's approach tends to overemphasize the individual's contribution to these processes,[18] this very recognition of human agency is crucial in the current reappraisal of his philosophy, which particularly emphasizes its performative orientation.[19] In Cassirer's framework, however, performativity is not defined as a question of how to do things with words; rather, the question is how to do things with a plurality of symbolic forms which here – in the topical essay from 1930 – also includes technology.

Agency and the formative effects of cultural tools are among the important aspects considered by Bertolt Brecht and Ernst Cassirer when they reflect upon the human possibilities of making use of modern technology and developing it in a productive way. Although Brecht and Cassirer worked in very different institutional contexts and neither of them engaged directly in the other one's works and ideas, they respond to challenges in their cultural environment in ways which supplement and confront each other – and which can contribute to contemporary discussions in the field of cultural studies. Thus, three of the notions dealt with in my previous reflections on post-World War II approaches to voice phenomena and listening reception will provide the basis for structuring the juxtaposing of the two thinkers in the next part of this chapter: their approaches to technology; to human expressivity; and to the impressionability of the subject.

Brecht versus Cassirer

Approaching technology

The radio is undoubtedly one of the technologies Ernst Cassirer *could have* analysed if his aim had been to develop an explicit and thorough reflection on one of the transforming tools of his own time. But this is not his focus in the essay 'Form and Technology'. Cassirer's reflections on technology are determined by an ambition to contribute to the integration of technology 'within the circle of philosophical self reflection',[20] and he locates technology among traditional craft tools. Furthermore, he stresses the shared origin of language and technology, describing how they both stem from a 'common root of forming gestalts, gradually unfolding and branching off from it'.[21] Yet, his theoretical reflections on technology are surrounded by the new mass media in more than one sense. John Krois has pointed out how Cassirer's own family was involved in the engineering and production of radio technology,[22] and in the volume *Kunst und Technik*, where 'Form and Technology' was first published, there are several contributions on the broadcast and its function in the contemporary culture of the Weimar Republic.[23] Leo Kestenberg, the editor of *Kunst und Technik*, argues, as a matter of fact, that these essays should be taken to supplement Cassirer's essay. According to Kestenberg's editorial preface, Cassirer's essay belongs to the philosophical and sociological contributions located at the very beginning of the volume, whose function is to provide 'a reliable foundation' for the following contributions which will then penetrate into the middle of the 'separate appearances'.[24]

In comparison, when Bertolt Brecht's scattered reflections on the radio have frequently been referred to as his 'radio-theory', this must be said to amount to something of an exaggeration. However, Brecht's texts do contain an explicit reflection on the new mass medium and its political implications in the late Weimar Republic, the cultural environment in which both Brecht and Cassirer lived and worked until they both had to flee from the Nazi regime in 1933. The radio needs to be 'refunctionalized' ('umfunktioniert'), Brecht claims; it must be transformed from a 'distribution apparatus into a communications apparatus'.[25] The quotation is from one of his speeches in the early 1930s, and his commentary goes beyond the mere ambition of producing avant-garde works which involve and engage the listening audience in new ways. At this point, Brecht was well aware of the technological and political origin of the limitations which determined the development of the German radio medium. Whereas American radio history started with radiotelephony, with hundreds of amateur and semi-commercial broadcasters, amateur broadcasting was prohibited in the early Weimar Republic. The state control and censorship which characterized German broadcasting from its inception was supposed to prevent political interference in radio programmes – an idea based upon an understanding of the state as an institution elevated above the political sphere. Despite these restrictions, the number of radio listeners increased rapidly and, by 1930, Germany had 3 million radio owners.[26]

According to the general expectations which determined German radio policy during this period, the new radio medium should be apolitical, contribute to the education of the people and provide the population with advice on how to cope with the challenges of modern life. Such issues are also reflected in *Kunst und Technik*, in the volume where Cassirer's essay was first published. One of the contributions dealing with the radio presents the organizational structure of the German broadcasting institution, describing the functions of its numerous boards and directors. Its author, Kurt von Boeckmann, emphasizes the federal legitimacy of this organization, and stresses how, from its very beginnings, German radio has fortunately been 'detached', keeping a safe distance from pure commercial interests, from the one-dimensional interests of political parties and from amateurism.[27] Another essay, written by Ernst Hardt, elaborates on the radio medium as part of a vital oral tradition with the potential to unify, to bring together different groups in society: 'scholars, poets and the people'.[28] Both contributions reflect the national radio policy of the period – Hardt's support for the law prohibiting the broadcasting of parliamentary discussions perhaps most strikingly so. Such mass mediation of political negotiations cannot be recommended, he argues, because the participants in the debates, the politicians, are not yet mature enough for this kind of public exposure.[29] Again, the implementation of the new technology is evaluated according to its expected impact on democratic development.

Brecht takes a different perspective in his texts on the radio medium: when the politicians are afraid of their voices being heard throughout the entire country, their fear is very legitimate, he writes. But their fear should not be allowed to determine broadcasting policy. In his 'Suggestions for the Director of Radio Broadcasting' (1927), published in the *Berliner Börsen-Courier*, he therefore recommends closer contact with the political negotiations, and suggests broadcasting directly from important public events and from institutional life in the Weimar Republic: '*I believe that you must move with the apparatuses closer to the real events and not simply limit yourself to reproducing or reporting.* You must go to the *parliamentary sessions* of the Reichstag and especially to the *court trials.*'[30] At the same time, Brecht claims that a transformation is needed of the very technology, in order to make real 'exchange' possible between the different parties, rather than simply facilitating distribution from one party to another.[31]

Compared to Brecht's reflections on the social effects of the new medium, a far more affirmative commitment to the radio as a democratic and liberating tool is traceable in the works of other commentators. For example, Leo Kestenberg claims in his preface to *Kunst und Technik* that, whereas art remains the empire of the few, technology is everybody's empire.[32] The philosopher Hans Reichenbach, one of Cassirer's friends, argues along similar lines, but with a particular emphasis on the radio as *the* cultural resource. In the handbook *Was ist Radio* (1929), Reichenbach identifies the radio medium as proof of how the increasing power of technology is able to bring 'intellectual wealth to all levels of society', while simultaneously mocking the 'ridiculous fear of the technologisation of culture'.[33] The virtue of the new medium is here identified as its capacity to distribute culturally valuable content to a larger segment of the population. According to both Cassirer and Brecht, however, technology cannot simply be regarded as a neutral tool for distribution; it is always also a formative tool which shapes its contents and its users.

Cassirer's approach to technology is more affirmative than that of Brecht. Moreover, the focus of his affirmation is on a different level and has a different function. And, unlike Reichenbach, Cassirer's aim is not to identify positive *effects* of technology; rather, it is to avoid an alienating effect being ascribed to technology as such. Both in 'Form and Technology' and elsewhere in his oeuvre, Cassirer explicitly refuses to commit to any kind of general scepticism which claims that culture, language or technology are alienating forces in themselves. He frequently clarifies in what respects his position differs by juxtaposing it to Georg Simmel's version of the 'Tragedy of Culture', according to which modern man's faith is negatively determined by the fact that he produces a variety of cultural artefacts without being able to integrate them into his world. The result of this production is that his free subjectivity is weakened. Since he is unable to penetrate or transform these artefacts with his own spirit, his original life spirit ebbs away. Cassirer agrees with Simmel that such a tragic aspect of cultural development can indeed be observed in modern society, and is perhaps nowhere 'more evident than in the development of modern technology'.[34] But, and this is a crucial point for Cassirer, such a development cannot serve as an argument against modern technology as such.

Unlike Simmel – and with a relevance which goes far beyond Cassirer's negotiations with him and addresses topical issues in contemporary approaches to mediation – Cassirer is critical of theoretical approaches which totalize their scepticism. Whereas Cassirer's position has been criticized as too 'optimistic', particularly with respect to the capacity of technology to contribute to human 'self-liberation',[35] Birgit Recki differentiates between a metaphysical and a 'practical optimism', pointing out that Cassirer's optimism is of the second type. Hence, Recki claims, his affirmative stance has to be considered equal to a 'positive working hypothesis'.[36] This is an insight which takes into account Cassirer's dynamic and pragmatic understanding of concepts and ideas as tools for systematic reflection *and* for developing democracy. The clearest expression of the latter issue is probably found in a text written by Cassirer in 1928 and given as a public speech the following year during a university celebration of the 1919 Weimar constitution. In this text he presents an historical examination of the idea on which the republican constitution is based, concluding that this very idea should be directed towards the future, a future which needs to be 'heraufgeführt', to be produced.[37] A similar idea-based, future-oriented pragmatism, nurtured by vital impulses from the tradition of enlightenment, determines his approach to technology in the essay from 1930.

Expressive meaning

When Brecht elaborates on his own radio play *The Flight of the Lindberghs* (*Der Flug der Lindberghs*) (1927), his premise seems to be that the expressivity of the radio-mediated voice tends to create listener identification rather than make the listener reflect critically on the mediated content. He emphasizes how the role of the hero ought to be sung by a choir in order to create the appropriate distance and avoid having the listener identify with the hero: 'Only collective I-singing ... can salvage something of the pedagogical effect,'[38] Brecht argues. Again, his critical approach to the radio-mediated voice and its capacity to make the listener identify with the protagonist is inseparably tied to reflections on cultural participation and democracy. Similar to the tendencies seen in McLuhan's and Bal's approaches, Brecht suggests that auditive reception is a situation in which the listener can easily be deprived of his capacity to understand and learn more, due to the lack of distance and clarity produced by the very mode of presentation.

Brecht's work on the Lindbergh play belongs to the period of his experimental learning plays ('Lehrstücke'). A photograph from the 1929 music festival in Baden-Baden (Figure 7.1) shows how his radio experiment was performed on stage with placards naming both the listener and the new technological tool, as if to make the new technology and the new genre of the radio play comprehensible through an exposition highlighting the separate roles of the apparatus and the listener.

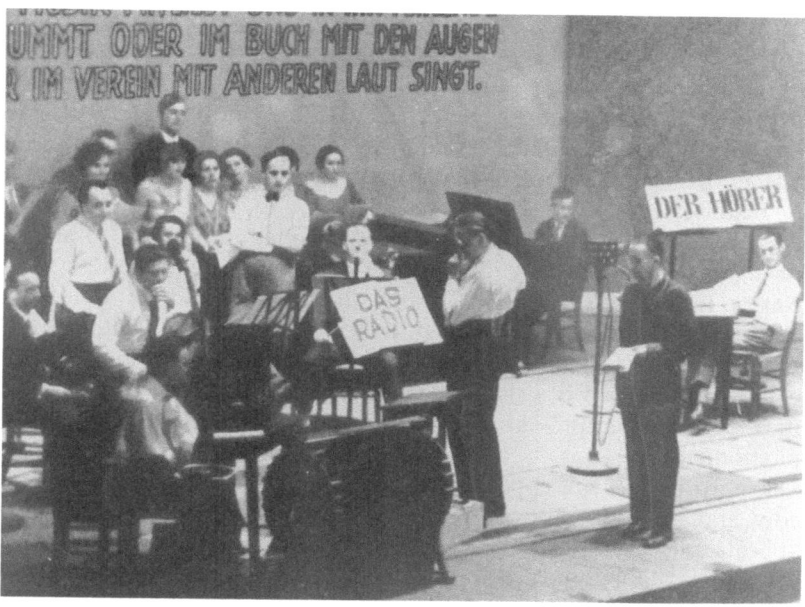

Figure 7.1   Brecht's radio experiment performed on stage, at the music festival in Baden-Baden, 1929

From a contemporary perspective, this staging of the play in Baden-Baden could almost be seen as a didactic anticipation of McLuhan's radio scepticism, with its reference to the mythical dimension and the intimate person-to-person aspect involved in radio reception. In this light, the staging blocks any intimate affective involvement with the radio. Instead it makes visible the parties involved – as separate agents. As a supporting supplement to the voice mediation, it delivers text, a script, readable for McLuhan's allegedly 'neutral eye'.[39] In short, it stabilizes and demythifies by creating an explanatory model. It brings its audience of potential radio listeners into the position of the critical observer, makes the new technological tool an object of critical scrutiny. The photo visualizes another important dimension in Brecht's experimental work during this period: the role of the radio is inhabited by a group of people, thus reflecting Brecht's focus on collective practices and collaborative productivity.

Whereas Brecht focused on how to create distance to the powerful expressivity of personal voice, Cassirer sought to integrate expressivity as an indispensable level of meaning into his system of thought. In his *Philosophy of Symbolic Forms* Cassirer describes such an expressive level of meaning as the 'living efficacy'[40] to which man is subordinated – but in which he is also actively involved, with his emotions, imagination and all of his senses. Thus, in contrast to Bal's analysis, the powerful effects of voice phenomena such as the 'personifications' manifested in historical radio plays from the 1930s, or in contemporary multimedia works of art such as Coleman's installation referred to above, need not be devalued as distractions from 'clarity of vision'.[41] They can be framed, instead, as belonging to the 'fundamental act of the mythical consciousness',[42] through which an 'original consolidation of emotionality'[43] takes place. This very recognition of an indispensable expressive function is a significant aspect when Cassirer's approach to meaning has been proven valuable in recent historical studies of voice phenomena.[44]

The world does not appear to us as 'stuff and matter', Cassirer repeatedly argued, addressing the empiricists of his own time. Rather, it appears to us as unified phenomena of expression to which we respond with our entire bodily and spiritual resources from our position within culture, surrounded by and shaped by culture. Without losing its relevance, his argument could easily be transferred to the current environment of literature and culture studies: the world does not appear to us as signs or codes, but as unified phenomena of expression, to which we respond actively. However, this should not be taken to mean that Cassirer celebrates expressivity and dismisses the function of the sign. What he offers is a framework in which the function of the sign rests upon a level of expression in which there is no clear distinction between sign and meaning, subject and object. The different symbolic forms, then, are formative, in the sense that 'each of them designates a different approach, in which and through which it constitutes its own aspect of "reality"'.[45]

In 'Form and Technology', Cassirer describes the dynamics of myth and technology as processes of differentiation and mutual confrontation, developing his argument around the German term 'Auseinandersetzung':

[T]he relation between both [between 'I' and 'reality'] is not set down as unique and unambiguous from the beginning. It first comes to be because of the manifold ideal processes of 'mutual differentiation and determination', as in myth and religion, language and art, science and the different basic forms of 'theoretical' conduct in general. For human beings, a fixed relation of subject and object according to which they conduct themselves does not exist from the beginning. Rather, in the entirety of a human being's activity, in the entirety of his bodily and his psycho-spiritual activities, there first arises knowledge of both subject and object; the horizon of the 'I' first separates itself from that of reality. There is no solid, static relation between them from the outset. Rather there is, as it were, a fluctuating movement of back and forth. From this movement a form gradually crystallizes in which the human being first grasps his own being as well as the being of objects.[46]

Time and again, Cassirer uses the concept of 'Auseinandersetzung' in order to describe the cultural processes through which meaning is developed – translated as 'mutual differentiation and determination' in the English translation which constitutes the first part of this volume. The term itself represents a true challenge for anyone translating Cassirer's texts into another language. In everyday usage, 'Auseinandersetzung' can simply mean 'struggle', 'debate', 'quarrel', thus pointing towards different modes of verbal interaction between opposing parties. In Cassirer's dynamic approach to cultural work, however, 'Auseinandersetzung' is coined as a phenomenological concept describing physically and spiritually determined processes of oscillation and mutual determination through which more stable forms can gradually emerge.

Furthermore, it is crucial to recognize how Cassirer's understanding of meaning as continuous cultural work evolves around the notion of embeddedness. From a position that is always already involved in and determined by a world of cultural utterances and images, meaning is formed through processes of separation and juxtaposition, and developed through active human intervention. The more stable and fixed forms of meaning resulting from these processes are hence conceptualized as forms which are both rooted in and part of a proximate involvement with the world. They are literally worked *out* and *set* apart, as units or structures confronting and mutually determining one another. As a consequence, production of meaning is not a matter of *representing* an already existing world; rather, it is a process of gradual development and exposition to which not only language, but also other significant forms such as technology and theory contribute productively.

However, Cassirer's recognition of expressivity as an indispensable level of meaning is always counterbalanced by an equally strong commitment to systematic comparison. Furthermore, his integration of myth as meaningful expressivity appears to be motivated not primarily by an interest in myth for its own sake, but by the 'need of knowledge and cognition'.[47] Consequently, as Cassirer phrases it in the introduction to the second volume of his *Philosophy of Symbolic Forms*, '[k]nowledge does not master myth by banishing it from its confines'.[48] Myth must be integrated – in order to maintain the unity of knowledge – and it must be mastered, according to an understanding of human development and emancipation as the self-liberation of thought from its most subjective foundations. In the context of our juxtaposition of Cassirer and Brecht, however, it is possible to trace how Cassirer's increased focus on the fact that cultural development does not automatically lead to emancipation – that modern mythical powers can threaten humanity at its very foundations – brings the Marxist author and the liberal philosopher closer to each other. This issue becomes particularly evident when we take into account Cassirer's posthumously published work *The Myth of the State* (1946).

The open subject

On the one hand, we have Cassirer's essay from 1930, which invests in the notion of an open and enlightened human subject committed to the recognition and cultivation of his environment through the use of language, tools and technologies; on the other, we have Brecht's far more practical and critical approach to the radio, which excludes a similar openness, at least for the time being. Instead of the paradigm of friendship and reciprocity which is implicit in Cassirer's idea of dynamic and mutual determination discussed above, Brecht places the subject in a far more violent field of conflicting interests. In the drafts for his play *Man Equals Man* (*Mann ist Mann*, broadcast in 1927), Brecht draws attention to the capacity of technology to fundamentally change humanity. The new technologies will contribute to the development of '*a new type of human being*',[49] Brecht argues, and this 'new type' is not going to 'let himself be changed by the machines; he is going to change the machines'.[50]

Both technology and man are here presented as modifiable entities, and even though Brecht and Cassirer share this idea of a human being who is dynamically shaped by his use of tools, the options presented by Brecht are different: man can either gain access to and control over technology, or he can be controlled by it and thus be changed by a powerful alien force. Both possibilities, however, involve a transformation of what it is to be a human being. When Brecht's main character Galy Galy, in *Man Equals Man*, gains strength, he does so by giving up his individuality and becoming part of the mass.

It is only in *The Myth of the State* that Cassirer spells out a similar option: a cultural development which questions the entire foundations of the tradition upon which he has based his system of thought. Here, there is a far stronger emphasis than in his earlier writings on the necessity of structuring the world of myth into an ordered universe. This change of emphasis becomes particularly evident in the last part of the text, where Cassirer considers how the manipulation of myth has been made possible by 'skilful use of new technical tools'.[51] His critical examination of his own conception of myth here forms part of a retrospective analysis of the propaganda machinery of the Third Reich. According to Cassirer, the new situation from which he has to re-evaluate his own approach is one in which the totalitarian regime has demonstrated its ability to manufacture myth – in the same sense and according to the same methods that it has used to manufacture its other weapons, such as machine guns or aeroplanes. This politically manufactured myth is the force which has demonstrated its capability to change the entire form of our social life.

In this part of his oeuvre, Cassirer describes a transformation of man in terms which are more compatible with Brecht's critical perspectives. He refers to a 'mental rearmament'[52] brought about by the politically manufactured myths – entailing unprecedented possibilities to manipulate man and his perception of himself and the world. Cassirer's positive recognition of the fluidity of the mythical – its plasticity – is here replaced by a notion of myth as that which exposes man to manipulation, similar to the tenor in the contemporary approaches dealt with in the first part of this chapter. It is exactly here, where the world is not fixed and determined, that it can be shaped by a violent force. At this point, Cassirer's conception of myth is determined by an ambivalence similar to the one expressed by Welsch in his reflections on a possible development towards an 'auditive culture', referred to above. In 'critical moments of man's social life', Cassirer writes, 'the rational forces that resist the rise of the old mythical conceptions are no longer sure of themselves. In these moments the time for myth has come again. For myth has not been really vanquished and subjugated. It is always there, lurking in the dark and waiting for its hour and opportunity.'[53]

Confronted with the experience of the totalitarian state, Cassirer warns against a perverted version of myth, a myth which comes 'again'.[54] Experiencing the powerful manipulative force of the National Socialistic regime and the frailty of the binding forces in society obviously challenges Cassirer's previous conceptualization of transformation processes leading from a mythical level to the level of more stable structures of meaning. Man can be manipulated, modern technologies are able to reach him in new ways; they are even capable of blocking his possibilities to develop his potential. In *The Myth of the State* it is the fusion of myth and technology which blocks the necessary interplay between the different symbolic forms.

Crucially, and this is a main point for Cassirer, this is a *possible* development; it is not one that will *necessarily* occur. And, unlike Brecht's reflections in *Man Equals Man*, for Cassirer, giving up individuality in favour of a strong mass identity can never serve as the alternative to technological manipulation of man. When, in this posthumously published text, Cassirer analyses the development that led to the Third Reich, he includes many of the well-known social historical aspects. It is worth noticing, however – and related to our earlier observation of how Cassirer includes theory among the symbolic forms – how the philosophers Martin Heidegger and Oswald Spengler are dealt with in the final part of *The Myth of the State*. Although Cassirer underlines that he is not suggesting that their philosophical doctrine had a 'direct bearing on the development of political ideas in Germany',[55] he nonetheless claims that they served as pliable instruments for the political leaders, because, after all, they 'did enfeeble and slowly undermine the forces that could have resisted the modern political myth'.[56] Again, Cassirer emphasizes theory as one of the cultural tools with which we shape our world. Philosophy, too, has its share in the 'construction and reconstruction of man's cultural life'.[57]

In this aspect, and particularly with regard to his above-mentioned emphasis on the social effects of theory, Cassirer's reflections from 1946 have significant traits in common with Rancière's contemporary examination of the 'misadventures of critical thought',[58] as pointed out in the first part of this chapter. The obvious differences taken into account – Cassirer is addressing Spengler and Heidegger, whereas Rancière examines a critical theory that has become doxa – Cassirer and Rancière share not only an awareness of the formative power of theory, but also a capability to invest in open situations: the idea that 'every situation can be cracked open from the inside, reconfigured in a different regime of perception and signification'.[59] The 'inside' should here not be conceived of as some kind of privileged innerness, but rather as the embodied and situated subjectivity through which knowledge can be developed.

Brecht and Cassirer: addressing posterity?

Whereas Brecht's radio reflections have been overestimated in that they are taken to present a 'radio theory', Cassirer's theoretical essay

'Form and Technology' has received far less scholarly attention – as pointed out in the introduction to this volume. However, this is about to change. Cassirer's philosophy addresses 'posterity' in the sense that it enables contemporary scholars to frame their fields of investigation in new ways. Its particular relevance for the investigation of voice phenomena and listening reception lies in Cassirer's recognition of a plurality of symbolic forms which includes technology and theory, and in his recognition of expressivity as an integrated and indispensable part of the human production of meaning.

The juxtaposition of Cassirer and Brecht draws attention to obvious differences – and to their shared concern about the development of democracy in a politically polarized cultural environment. Furthermore, Cassirer and Brecht confront each other's views productively. Brecht's strong concern for the institutional developments of the late Weimar Republic, for instance, focuses on an aspect which is certainly underestimated in Cassirer's approach to meaning during the 1920s and the 30s. Interestingly, however, the fact that Cassirer underscores individual agency strikes me today, somehow, as a corrective; or, more correctly, as a reminder of a cultural agent almost forgotten after a period of powerful theoretical conceptualizations of the subject as the mere product of social structures, epistemes and discourses.

With a Cassirerian approach to cultural meaning it is possible to reflect on the Brechtian idea of the V-effect, artistic strategies of distancing in a more relative, dynamic and historical way which takes us beyond the neat separation of sensations and decisions, feeling and reason.[60] The basis for such reflections would then be an approach to meaning which is – as pointed out above – neither based upon the notion of representation nor upon the notion of the sign. Signification does not *create* meaning; it merely 'stabilizes' an already existing meaning, Cassirer writes.[61] In its original context, this emphasis on the stabilizing function of language appears as an explication of his seminal theoretical point, namely that the function of language is not to repeat already existing definitions and distinctions; rather, it is to formulate them, and to make them intelligible. In our context, of early German radio history, this emphasis upon the stabilizing function of signification appears as interesting and thought-provoking in yet another sense. When Brecht put the radio on stage in the late 1920s, creating a didactic model which explained, with posters and placards, how the new technology should work, the stabilizing intention was obvious, and so was the normative and didactic impulse in his radio reflections from this period. And yet, Brecht's approach differs clearly from the much more radicalized media

scepticism from the second part of the twentieth century. His reflections on mediation have a far more practical basis. He experiments
and reflects on the new mass media – with the aim of contributing to
media literacy and social change in a specific situation in the history of
the media. Both in his more explicit systematic reflections and in his
interesting short radio poems, however, he recognizes human expressivity as meaningful *and* highly problematic. In this sense it is possible to
elaborate further on his ideas on how to achieve a productive distance
as part of a more dynamic critical practice; as a practice which acts on,
and thereby presupposes, the level of meaning which is conceptualized
by Cassirer as expressive meaning, as 'Ausdruckswahrnehmung'.

In this sense my reading of Ernst Cassirer has taken as its starting point
a contemporary situation in which the conceptualization of voice phenomena and listening reception needs to be developed with a stronger
systematic recognition of expressive meaning as genuinely meaningful.
Furthermore, I have pointed out as a particular resource in Cassirer's
approach to myth and technology his ability to theorize on cultural crises without turning the state of exception into universal law. Although
the symbolic capacity of man can be blocked and limited, Cassirer
retains the possible options of productive transformation, cultural development and the integration of new tools in the human sphere. One of
the most powerful articulations in Bertolt Brecht's oeuvre of *his* reflexive
awareness of his own role as an influential author strongly determined
by times of crises is found in a draft version of the poem quoted at the
start of this chapter. Here, as if to warn against any reading practice
which canonizes certain aesthetic strategies, particular modes of writing,
without taking into account their cultural context and their structure
of address, Brecht turns towards his future readers with an explicit con-
textualization of his own writing: 'You, future generations, when you
read the things I wrote / Consider, friends, the time I wrote them in. /
Whatever you may choose to think, do not forget / This time.'[62]

Notes

1. The recorded reading of the poem is available on the CD *Bertolt Brecht – An
die Nachgeborenen* (München: Der HörVerlag, 1997); the written poem in
Bertolt Brecht, *Werke*, ed. Werner Hecht, Jan Knopf, Werner Mittenzwei
and Klaus-Detlef Müller (Berlin: Aufbau, 1988), vol. 12, 85ff. The English
translation is quoted from Bertolt Brecht, *Selected Poems*, trans. H. R. Hays
(New York: Harcourt, 1947), 177.
2. Winfried B. Lerg, *Die Entstehung des Rundfunks in Deutschland* (Frankfurt
am Main: Josef Knecht,1965), 310. The English version of the quotation is

my own translation. Likewise, below, all quotations from texts written in German but rendered in English and not accompanied by a reference to an English translation of the source are my own translations.

3. Marshall McLuhan, *Understanding Media: The Extensions of Man* (Padstow: T. J. Press, 1995), 299.

4. *Ibid.*

5. *Ibid.*, 301.

6. Wolfgang Welsch, *Undoing Aesthetics*, trans. Andrew Inkpin (London: Sage, 1997), 151.

7. Welsch, *Undoing Aesthetics*, 157.

8. *Ibid.*, 158.

9. Mieke Bal, *Travelling Concepts in the Humanities* (Toronto: University of Toronto Press, 2002), 174–208.

10. *Ibid.*, 184.

11. *Ibid.*, 192.

12. *Ibid.*, 191.

13. Rudolf Arnheim, *Rundfunk als Hörkunst* (Frankfurt am Main: Suhrkamp Verlag, 2001), 92.

14. Friedrich Kittler, *Gramophone, Film, Typewriter*, trans. Geoffrey Winthrop-Young and Michael Wutz (Stanford: Stanford University Press, 1999), 10.

15. Jean Rancière, *The Emancipated Spectator*, trans. Gregory Elliott (London: Verso, 2009), 42.

16. See Cassirer's 'Form and Technology' in this volume, 34.

17. *Ibid.*, 24.

18. On this topic, see Esther Oluffa Pedersen's critical discussion of Cassirer's and Usener's hypotheses on the 'originäre Prädikation', in *Die Mythosphilosophie Ernst Cassirers* (Würzburg: Königshausen and Neumann, 2009), 172f.

19. See Barbara Naumann and Birgit Recki, 'Einleitung', in *Cassirer und Goethe*, ed. Naumann and Recki (Berlin: Akademie Verlag, 2002), IX.

20. Cassirer, 'Form and Technology', 17.

21. *Ibid.*, 24.

22. See John Krois' article in this volume, 'The Age of Complete Mechanisation', 56.

23. See *Kunst und Technik*, ed. Leo Kestenberg (Berlin: Wegweiser, 1930).

24. *Ibid.*, 11.

25. Bertolt Brecht, *Brecht on Film and Radio*, trans. and ed. Marc Silberman (London: Methuen, 2000), 42.

26. Carsten Lenk, *Die Erscheinung des Rundfunks. Einführung und Nutzung eines neuen Mediums* (Opladen: Westdeutscher Verlag, 1997), 125.

27. Kurt von Boeckmann, 'Organisation des deutschen Rundfunks', in *Kunst und Technik*, 222.

28. Ernst Hardt, 'Wort und Rundfunk', in *Kunst und Technik*, 180.

29. *Ibid.*, 181.

30. Brecht, *Brecht on Film and Radio*, 35, original emphasis.

31. *Ibid.*, 37.

32. *Kunst und Technik*, 8.

33. See Hans Reichenbach and Fritz Noack, *Was ist Radio* (Berlin: Richard Carl Schmidt & Co., 1929), 139.

34. Cassirer, 'Form and Technology', 41.

35. See, for example, Dirk Lüddecke, *Staat – Mythos – Politik: Überlegungen zum politischen Denken bei Ernst Cassirer* (Würzburg: Ergon, 2002), 366f.
36. Birgit Recki, *Kultur als Praxis: Eine Einführung in Ernst Cassirers Philosophie der symbolischen Formen* (Berlin: Akademie Verlag, 2004), 177.
37. Ernst Cassirer, 'Die Idee der republikanischen Verfassung: Rede zur Verfassungsfeier am 11. August 1928 (1929)', in Ernst Cassirer, *Gesammelte Werke*, ed. Birgit Recki (Hamburg: Meiner, 1998–2009), vol. 17, 307.
38. Brecht, *Brecht on Film and Radio*, 44.
39. McLuhan, *Understanding Media*, 302.
40. Ernst Cassirer, *Philosophy of Symbolic Forms*, vol. 3: *The Phenomenology of Knowledge*, trans. Ralph Manheim (New Haven: Yale University Press, 1957), 73.
41. Bal, *Travelling Concepts*, 172.
42. Cassirer, *Philosophy of Symbolic Forms*, vol. 3, 70.
43. Pedersen, *Die Mythosphilosophie Ernst Cassirers*, 199f.
44. See Reinhardt Meyer-Kalkus, *Stimme und Wahrnehmungskünste im 20. Jahrhundert* (Berlin: Akademie Verlag, 2001).
45. Ernst Cassirer, *Philosophy of Symbolic Forms*, vol. 1: *Language*, trans. by Ralph Manheim (New Haven: Yale University Press, 1953), 78.
46. Cassirer, 'Form and Technology', 26.
47. Ernst Cassirer, *Philosophy of Symbolic Forms*, vol. 2, *Mythical Thought*, trans. Ralph Manheim (New Haven: Yale University Press, 1955), xvii.
48. *Ibid.*
49. Brecht, *Werke*, vol. 24, 40, original emphasis.
50. *Ibid.*, 41.
51. Ernst Cassirer, *The Myth of the State* (New Haven: Yale University Press, 1946), 276.
52. *Ibid.*, 282.
53. *Ibid.*, 280.
54. As pointed out by Birgit Recki, this version of myth has strong similarities to Adorno's reflections on a 'second mythology'. See Recki, *Kultur als Praxis*, 107.
55. Cassirer, *The Myth of the State*, 280.
56. *Ibid.*, 293.
57. *Ibid.*
58. Rancière, *The Emancipated Spectator*, 25ff.
59. *Ibid.*, 49.
60. See, for example, Bertolt Brecht, 'The Modern Theatre is the Epic Theatre', in Brecht, *Brecht on Theatre*, trans. John Willett (London: Methuen, 1964), 37.
61. Cassirer, *Philosophy of Symbolic Forms*, vol. 1, 107.
62. 'Ihr Nachgeborenen, wenn ihr lest, was ich schrieb / bedenkt auch, Freundliche, die Zeit, in der ich schrieb. / Was immer ihr denken möget, vergeßt nicht / diese Zeit', in Brecht, *Werke*, vol. 12, 358.

8
'Representation' and 'Presence' in the Philosophy of Ernst Cassirer

Marion Lauschke

If we wish to read Ernst Cassirer from a contemporary perspective and want to ascertain his philosophical significance, it might make some sense to follow how Foucault fared in his similar attempt with Hegel. In his inaugural address at the Collège de France – later published under the title *The Order of Discourse* – Foucault asserts that it is never easy to distance oneself from Hegel: 'In order to really free oneself from Hegel, we first have to assess the cost of renouncing him. We have to realize the extent to which Hegel perhaps secretly influences us; that our thoughts against him might actually come from him.'[1] Of course Cassirer isn't Hegel, and we can't characterize the 65-odd years that separate us from Cassirer as a time of opposition against him, since Cassirer has been either ignored or forgotten for two-thirds of those 65 years.

But if we can follow the many 'turns' of contemporary cultural studies, ignorance of Cassirer appears to have resulted in a multitude of distracting detours. It is easy to picture Cassirer, considered antiquated even in his own lifetime, calmly watching the twist and turns of contemporary discourse and thinking to himself, 'I have already been there.' Indeed, one of Cassirer's idiosyncrasies, contributing to his reputation as a stick-in-the-mud, was his respect for the philosophical tradition. He was not a thinker of big gestures, nor did he coin new concepts unnecessarily, though his subtle variations in meaning often made these traditional concepts appear in a whole new light.

The Theory of Representation, dismissed by Foucault as a relic of the nineteenth century, belongs to the Cassirean revising of thought. The theory, though even then acutely in crisis, clung stubbornly to life. It may be hard to understand why Cassirer did not abandon this contested concept of representation, given that his symbolic forms do not represent actual things or facts of some originally given world. Cassirer was well acquainted with the history of the concept of representation and with the many critiques of it. And, as a Kantian, he was unshakable in his opposition to any attempt at Metaphysical Realism. He was an outspoken critic of a certain kind of Representation Theory, namely Dual Representation Theory, which he always criticized as a form of mental copy theory.

In the following chapter, I will present Cassirer as an advocate of a constructivist theory of representation and will try to get to the bottom of the relation between 'representation' and 'presence' (or 'presentation') in the philosophy of symbolic forms.[2] By 'representation', Cassirer means a relation internal to consciousness. It is not a substitute for a primary presence, but rather its precondition, and is closely related to the conception of symbolic forms and to the symbolic pregnance of perception.

This chapter contains three sections. In the first section, I will differentiate the concept of 'representation' that underlies the philosophy of symbolic forms from the concept of 'presentation'. In the second section I will show to what extent the concepts 'representation' and 'presence' – and their fluctuating relation – help us to differentiate the symbolic forms from one another. From this perspective, I will then shift my focus in the third section towards art-aesthetic phenomena and discuss their unique status, characterized by an oscillation between 'presence' and 'representation'.

I

As a philosopher in the Kantian tradition, Ernst Cassirer's philosophy of symbolic forms follows in the footsteps of critical idealism, which 'renounces the proud name of ontology',[3] and contents itself with the modest task of analysing the formative forces of humankind. It understands 'reality' not as a unity, because both subject and object are first the result of the most manifold entanglements of human beings and their sensory impressions. According to this revolutionary way of thinking, there is no longer the one objective being capable of faithfully reproducing a thought. In dealing with its products, then, mind invariably relates to itself. By displacing the production sites of mental images out of the space between subject and object and moving them to the interior space of consciousness's forming processes – notice here the farewell of mental copy theory – the concept of image itself changes. The images human beings make of the world are now understood as 'virtual images'.[4] Cassirer labels all areas of symbolic formation as:

particular image-worlds, which do not merely reflect the empirically given, but which rather produce it in accordance with an independent principle. Each of these functions creates its own symbolic forms which, if not similar to the intellectual symbols, enjoy equal rank as products of human spirit. None of these forms can simply be reduced to, or derived from, the others; each of them designates a particular approach, in which and through which it constitutes its own aspect of 'reality'.[5]

For the symbolic cosmopolitan there are no archetypes that can be mimicked or original forms to trace other forms back to, just as there is no 'native language' in which other languages must be translated, because the human being, the 'symbolic animal', is at home in every symbolic 'language'. Symbolic forms are the results of continuous formation processes the human being initiates and continually updates in the most diverse ways and with the most diverse means. They are deprived of allegorical interpretation, because to allegorically understand an object for Cassirer means to understand it only 'by referring it and reducing it to something other than what it immediately is and signifies'.[6] Cassirer labels the philosophy of symbolic forms' point-of-view 'tautegorical'. Here that means understanding the respective production of images as autonomous mental creations, 'which one must understand from within by knowing the way in which they take on meaning and form'.[7]

Although subject and object do not lie ahead of the process of symbolic forming but first emerge from it (so that nothing appears 'given' that could be 're-presented' through a symbolic form), Cassirer maintains the concept of representation and uses it in two ways. First, following Leibniz, he understands representation as consciousness itself. Second, he understands representation to be the relation between a single phenomenon and symbolic form as a whole.

The concept of representation is tightly joined to the concept of symbol. Cassirer designates the function of representation as the foundation for the concept of symbol in Leibniz's system.[8] Already, in his early interpretation of Leibniz, Cassirer lays the groundwork for his later explanation that relations of conscious content are representations, because, along with the Leibnizian theory of monads, consciousness is understood as 'a unity in the manifold'.[9] It is not a substance, but rather represents the unity of changing conscious content. It is the very 'faculty' of the idea itself. Through his concept of aesthetic harmony, Cassirer attempts to explain 'the deeper meaning of the manifold of representation'.[10] The harmonious unity does not involve the additive relation of a 'one-to-one correspondence'. It is a unity preceding its individual parts. Cassirer follows Leibniz by emphasizing the activist connotation of 'exprimere' – of enunciation. Cassirer's interchangeable use of 'exprimere' and 'represents' is suggestive of how he understands the character of consciousness's capacity.

Cassirer does not limit his interpretation of representations to aesthetic perception and the whole consciousness. In *Substance and Function* Cassirer places every 'datum' into a representational dependency. He develops an understanding of representational relations that he then extends into the area of scientific knowledge. According to Cassirer, physical data first attain meaning through the 'multiplicity of their relationships' – through a datum's interaction with other data. The 'datum' becomes the 'symbol of a thorough systematic organization, within which it stands and to a certain extent participates'.[11]

Cassirer was familiar with the attacks levelled against the concept of representation throughout the history of philosophy. In his four-volume book, *The Problem of Knowledge*, he dealt with different forms of the subject-object construction that correspond to the concept of representation in the philosophical tradition and for which this notion was criticized. In *Substance and Function* Cassirer also defends the 'Concept of Reality' against sceptical arguments by alluding to the 'new meaning' given to the word in the critique of knowledge.[12] Transcendental Idealism, with which Cassirer allied himself, argues that 'real reality' – the 'thing in itself' – is unattainable. Cassirer's concept of representation is an internal relation where 'one element of consciousness is represented in and through another'.[13]

Cassirer uses the example of experimental results to explain general relations of representations. Experiments are only scientifically relevant if they unlock clues and thereby make a result possible. These clues obtained in the experiment are set in a complex correlation:

Each particular phase of experience has a 'representative' character, in so far as it refers to another and finally leads by progress according to rule to the totality of experience. But this reference beyond concerns only the transition from one particular serial member to the totality, to which it belongs, and to the universal rule governing this totality. The enlargement does not extend into a field that is absolutely beyond, but on the contrary, aims to grasp as a definite whole same field, of which the particular experience is a part. It places the individual in the system.[14]

A datum has a representative character within a sensory correlation and it represents exactly this correlation. Its relation to the 'outer world' is not nullified through an internal referencing, because its access to reality is the very quality of being able to represent such a correlation and the ability to make an expansion of such correlations possible. Indeed, this is humankind's only access to reality. It is impossible to abandon the world of image for the purpose of comparing images with 'real' objects. 'Real' objects are as much a fiction as the 'driving force' of human consciousness. It is the 'fertility' of an accepted representational relation that authenticates the real character of virtual images, and therefore allows Cassirer to understand the words of his cherished Goethe: the fruitful alone is true.[15]

In this general sense, we can ascertain the relation between representation and presentation in the following way: 'representation' means the 'ideal rule, which connects the present, given particular with the whole, and combines the two in an intellectual synthesis'.[16] For Cassirer, a representation is not the subsequent interpretation of something primarily given, but rather the condition of presentation – that from which humankind is able to become conscious of actual things. Cassirer follows the Kantian definition of apperception insofar as every sensory perception is determined by the conditions of the forms of perception and the order of the categories, but he liberalizes Kant by pluralizing the relations through which these ordered structures are established.

Without this 'apparent representation' – he writes 'apparent' because, in this aspect, he is not talking about a re-presentation, or repeated act from something original – 'there would also be no presentation, no immediately present content; for this latter only exists for knowledge in so far as it is brought into a system of relations, that give it spatial and temporal as well as conceptual determinateness'.[17] The special experience or insight therefore represents the whole of the respective context of experience.

The determination here obtained in the relation between presentation and representation by Cassirer is analogous to the determinations of the particular kinds of relations he already dealt with for classical concepts of opposition like 'form' and 'content' or the 'senses' and the 'understanding'.[18] Just as there is no such thing as unformed material for

Cassirer, there can also be no 'presence' in human consciousness that is not already 'representation'. Every attentive look at the world is, along with Goethe, already theory. 'It lies in the very nature of consciousness', Cassirer writes in the first volume of the *Philosophy of Symbolic Forms*, 'that it cannot posit any content without, by this simple act, positing a complex of other contents',[19] '[f]or what defines each particular content of consciousness is that in it the whole of consciousness is in some form posited and represented. Only in and through this *representation* does what we call the "presence" of the content become possible.'[20]

II

Besides this 'primal function of representation',[21] by which Cassirer means the structure or regular formation of the relationship between the general and the specific, and which is identical to the general 'symbolic function' or the 'natural symbolic'[22] of consciousness, Cassirer describes the relation of representation and presentation in the philosophy of symbolic forms as one which leads to the differentiation of symbolic forms. Although the condensed thesis regarding the concept of 'symbolic pregnance' – namely, that an elementary sensuous content is never 'there' as an isolated content, but rather is always formed as a 'concrete unity of "presence" and "representation"' – applies to all forms, Cassirer verifies that the 'relation of tension' between 'the content and representative function of a phenomenon ... does not stand out with equal fullness and distinctness in all the structures of consciousness'.[23]

To mark the characteristic differences, Cassirer differentiates the function of the symbol into its representational function, expressive function and signification. He then connects these functions with diverse forms of perception. Within this differentiation, the concepts of presentation and representation attain further meaning. They do not describe the relation between 'subject' and 'object' from the epistemological perspective of a Kantian; now, presentation and representation reconstruct an internal perspective within the subject, who constructs the symbolic forms to order life.[24]

In seeming contradiction to what I have written thus far regarding the representational character of consciousness, Cassirer differentiates between a 'mental' and a 'sensory consciousness'.[25] The 'sensory consciousness', or 'stratum of experience',[26] is characterized by a 'passive receptivity' to the 'chaos of sensory impressions'.[27] The 'world of sensation or intuition' of sensory consciousness follows the stream of

the imagination – the 'dream of images'[28] – because the 'basic character' of 'sensory consciousness' is the 'Heraclitian Flux'.[29] This level of
consciousness precedes, according to Cassirer, 'the divergence of myth
and theory, of logic, reflection and aesthetic intuition. Its certainty and
its "truth" are, in a manner of speaking, pre-mythical, pre-logical, and
pre-aesthetic; it forms the common ground from which all these formations have in some way sprung and to which they remain attached'.[30]
Expressive experiences or perceptions belong to this 'layer of experience'
that Cassirer calls an 'original mode of perception', or 'experiences of
pure expression [that] are not of mediated but of an original character'.[31]
Myth is rooted in these experiences of expression:

The mythical consciousness does not deduce essence from appearance, it possesses – it *has* – the essence in the appearance. The essence
does not recede behind the appearance but is manifested in it; it does
not cloak itself in the appearance given to itself. Here the phenomenon as it is given in any moment never has a character of mere
representation, it is one of authentic presence: here a reality is not
'actualized'– through the mediation of the phenomena but is present
in full actuality in the phenomenon.[32]

The fleeting moment of experience is tethered to the sensory present.
But this 'essence-having' is still only a fleeting kind of 'having'. It comes
undone as soon as another experience comes along. And while myth is
characterized by such phenomena of expression or presence, it does not
end there. Myth is also a structuring of the world. Through this presentation, the direct 'having of an experience', myth unites itself with
a representative purpose, and this representation belongs to the sphere
of 'judgement' – understood in the sense of an original structuring or
separation (literally, an Ur-Teilung).[33]

In order that the fleeting patterns of perception 'exist for us'[34] and
can be grasped beyond experiential time and become 'permanent and
stable',[35] they must be presented in the form of signs – in this sense they
must be *re*-presented. The part of the third volume of the *Philosophy of
Symbolic Forms* that deals with the problem of depiction – the second
part deals with the expressive function, the fourth with signification – is
called 'The Problem of Representation and the Construction of the
Perceptible World'. Every mental product that the philosophy of symbolic forms analyses is a representation, though they differentiate themselves through different 'relations of tension' between presentative and
representative moments.

While mythical and religious rites begin from the presence of the designated in the sign and a direct effect that results from contact with the 'sign' (not taken to be a sign of something else), the 'signs' of language obtain a representational character.[36] Language's depictions are 'symbolically used and symbolically understood'.[37] Even so, not all presentative phenomena are banished from language:

> All the phenomena that we call onomatopoeia belong to this sphere, for in the onomatopoeic formations of language we are dealing far less with the direct limitation of objectively given phenomena than with a linguistic formation which still remains wholly within the purely physiognomic world view. Here the sound attempts, as it were, to capture the immediate face of things and with it their true essence. Even where living language has long since learned to use the word as pure vehicle of thought it never wholly relinquishes this connection.[38]

Only upon the development of scientific concepts – and with them their corresponding sphere of meaning – did the 'representation function' achieve its full meaning. The characteristic 'attitude' of the concept exists, for Cassirer, in the fact that it 'must annul "presence" in order to arrive at "representation"'.[39] Myth is thereby identified as a predominantly presentative phenomenon, whereas language stands in a tension between 'presence' and 'representation', and the sciences are exclusively allocated to representation.

III

While the concept of representation has already come under fire through the critique of the copy theory of art and survives only in the concept of reference[40] that a hermeneutically oriented art and literature theory is not ready to relinquish, the discussion about 'presence' and 'presentation' remains contemporary. With the publication of Hans Ulrich Gumbrecht's *Production of Presence: What Meaning Cannot Convey* in 2004, this discussion has become even more relevant.

Derrida assumes that a sign, or that which is signified by the sign, can never be present or be an identity by itself. He formulates his poststructuralist concept of the sign as a critique of a 'Metaphysics of Presence'. Instead of representing already existing ideas, the meaning of a sign disperses into distinctions: 'There is not a single signified that escapes, even if recaptured, the play of signifying references that constitute language.'[41] The concept of performance, coined by a contemporary culture theory based on speech-act theory, is similarly anti-representational, though it still attempts to differentiate the actuality of, for example, a theatrical performance from an underlying (previous) text. 'In the mode of performance, a sign is what it means. Therefore, the material, medial, and the temporally enduring *presence* are the focus of attention.'[42]

In distinction to the wide-ranging deconstructive difference theories in art and literature theory discourses, there is a tendency in contemporary art to focus on 'presentation'. 'In contemporary forms of art,' claims Dieter Schlenstedt, 'where actions are understood as performances (in earlier dramatic art) or when materials or ready-mades are called art (in earlier fine art), or when the concrete poetry movement focused on the ramification of fragmented speech parts, it is clear that something should be identical with itself: the exhibiting human and the exhibited material. Representing is reduced to placing, or, more radically: "arrangement as presentation" opposes "portrayal as representation".'[43] Karl Heinz Bohrer has been investigating aesthetic phenomena for some time, and argues that they emerge from the 'suddenness' of aesthetic perception rather than the retroactivity of 'analytically processing, reflexive acts'.[44] Hans Ulrich Gumbrecht openly confesses to his 'longing for presence'[45] and attempts to rehabilitate the concept that has 'long been a symptom of despicably bad taste'.[46]

In this section of the chapter, I will introduce Cassirer's reflections into the discussion about 'presence' and 'representation' in art. On the one hand, Cassirer is a constructivist in the Kantian mould. His concept of representation does not share the epistemological naivety of copy theory. His concept of symbolic form does not expand into the negative infinity of the deconstructivist concept of signs. Instead, it rises with confidence in the production and relational processes of culture, by which it is shaped. The production and reception of the symbolic forms exist within a historical process in which understanding comes about through dialogue. On the other hand, his revaluation for the aesthetic side of the world, as well as the immemorial nature of the each symbolic world secures an image of humankind that safeguards him from the hubris of a builder of intellectual worlds.

Because of the receptivity of human sensibility, the human being is bound to the 'world' that he symbolically creates in a sensuous way. In the symbolic form of myth, the human being is at the mercy of his sensory impressions. In art, he freely lets himself be determined by these impressions. In art, the human being enjoys the sensory experience of attunement and takes back the 'to-be-in-the-world' feeling.[47] The concept of presence that Cassirer develops in the shadow of the concept of representation (which itself underlies the concept of symbolic forms) here obtains its meaning.[48] We can understand the aesthetic phenomenon in its oscillation between presence and representation.

In the first section, I showed that Cassirer conceived his symbolic forms as a further development of the Kantian concept of experience as coherencies that follow certain rules of connection. Every cultural fact is the result of a certain 'continuum of experience' and vice versa. In this sense, a myth represents the symbolic form of myth rather than the antecedent understanding of a human being living and thinking mythically. Similarly, a scientific fact represents the symbolic form of science and not the thoughts of the scientist. In accord with Derrida's critique of the 'Metaphysics of Presence', Cassirer does not believe the contents of consciousness to be autonomously contoured. Cassirer never describes a representation as a singular fact. It is 'never a substitute ... but the pertinent manner by which something moves from presentation and attains existence'.[49] He understands representation to be 'no mere subsequent determination, but a constitutive condition of all experience'.[50]

Works of art re-present the symbolic form of art. They can be understood as parts of the extensive continuum of form that develops during the history of literature, music and art. The literary concept of 'intertextuality' describes the special way that literature (and art in general) encodes itself into the existing symbolic form. Absorbing, transforming, or referring, literature continuously updates itself and develops deep historically specific dimensions of form. The notion that the sign has no reference, which appears to be the limit on the horizon of Derrida's theory of the sign, cannot, however, be united with Cassirer's symbol theory. For Cassirer, representations are relations within consciousness. But their 'worldliness' is demonstrated by the interaction of the 'I' with the world. Through this interaction, the subject is freed from the immediate entanglement of impressions. Following Paul Celan, symbolic forms are 'wounded by the real and in search of it'.[51]

As a representative of a symbolic form, a work of art stands for a perspective of the world. But every work of art is also an 'integral' world in itself and represents or exemplifies (following Nelson Goodman) itself. While science is to be understood exclusively as an extensive, advancing, series-forming representation, the representations of art are of a teleological or monadological nature. These relations – through which the parts of a work connect with each other – are individual and exist only once, in the single work in question.

The work of art as a whole represents the sense that develops out of these connections. Following Cassirer, Susanne Langer addressed the special symbolic form of art. 'The most original of Ernst Cassirer's contributions to philosophy, and perhaps the most important as well,' she writes, 'is his treatment of different forms of symbolic presentation and representation.'[52] The explicit difference between the two 'symbolic modes', discursive and presentational symbolic, is nevertheless original to Langer, who attributes art to presentational symbolic. The significance of presentational symbolic lies in the entire complex of 'established form' and cannot be analysed individually in contrast to the symbolization of discursive types. Artistic symbols give expression to specific qualities of emotional life that can only be symbolized through presentation. Especially through the conception of 'morphology of feelings', Langer further develops important general aspects of Cassirer's theory of perception, particularly his concept of art.

But we will not examine the concept of 'monadological' representation, which Langer allocated to presentation, because, besides the 'representational achievement of the symbol, which first establishes the phenomenon of conscious "presence"',[53] Cassirer himself deals with a form of presence that is prior to symbolic integration. Already in connection to the differentiation of the different symbolic forms in accordance with the relation of presence and representation, an 'original mode of perception' and 'experiences of pure expression [which] are not of a mediated but of an original character'[54] are apparent, which differentiates it from the 'presence through representation' characteristic of the symbolic forms.

In mythical consciousness, the phenomenon 'given in any moment never has a character of mere representation, it is one of authentic presence: here a reality is not actualized through the mediation of the phenomenon but is present in full actuality in the phenomenon'.[55] As a consequence of this analysis of mythical consciousness, I would like to follow Cassirer and differentiate 'between a presentative and a representative attitude of mind, between clinging to the sense impression and to sensuous objects',[56] and the transcending of this clinging into an ideal meaning. I would also like to apply this difference to aesthetic perception and the symbolic form of art. The presentative attitude, which 'builds the world out of images and adheres to this world',[57] is constitutive for both mythical consciousness and aesthetic perception.

Under the heading of 'sensory consciousness' Cassirer does mention a level of sensory perception that is not brought into some kind of connection with other sense impressions. Instead, the term designates the 'immediate being-there of the impression', a 'presentation' or a direct 'having'.[58] If the assumption of a pre-symbolic sensory consciousness appears to contradict the theory of symbolic pregnance, let us examine it in the context of aesthetic reflection as a question of changing mind-sets or mental attitudes:

> Instead of the dialectical movement of thought, in which every given particular is linked with other particulars in a series and thus ultimately subordinated to a general *law* and process, we have here a mere subjection to the impression itself and its momentary 'presence'. Consciousness is bound by its mere facticity.[59]

The mythical experience of expression – as well as the aesthetic perception itself – pertains to the 'transparence' of the sensory, which is incapable and is in no need of symbolic integration: 'The expressive meaning attaches to the perception itself, in which it is apprehended and immediately experienced.'[60] Aesthetic perception is directed towards sensory presence – 'the sphere of plastic, architectural, musical forms, of shapes and designs, melodies and rhythms' – and not towards representative meaning. 'As soon as I lose these sensuous forms from sight,' claims Cassirer, 'I lose the ground of my aesthetic experience.'[61] This applies to expressive perception in general, but especially to aesthetic perception: 'the simple baring of the phenomenon is at the same time its interpretation, the only one of which it is susceptible and needful'.[62]

A mental attitude of presence, one which does not hermeneutically transcend or exceed given phenomena but which remains bound to its presence and onto its sensory perception, does not necessitate a causal explanation, which would lead to a distancing from aesthetic phenomenon as such and could potentially lead to its disappearance:

> For a distinction of this sort would demand something more than *mere intuitive immersion in the content itself*; instead of apprehending the particular contents in their presence, the understanding would have to trace them back to the conditions of their genesis in consciousness and to the principle of causality governing this genesis.[63]

Leibniz had already examined this conscious commitment to 'appearance' or, to follow Goethe, to the 'aesthetic side of phenomena'. It is an original mode of perceiving sensuous presence. As a perceived immersion in the given singular, aesthetic perceptions are not linked to any extensive integral of experience. They are characterized by 'insularity', suddenness' or 'discontinuity'.[64] The 'life nerve of the temporal'[65] that develops from the integration of individual fragments is anaesthetized by the aesthetic perception. The moment of perception leaves the network of the temporal and stands still. Karl-Heinz Bohrer highlights the 'temporal modality of (the) "suddenness"' of aesthetic perception. Hans Ulrich Gumbrecht characterizes it primarily as a present experience of space, 'being-in-the-world' in which the subject feels the bodily presence of things. It is not an interpretive viewing of the world, but an experience of space characterized by a lack of distance. During an aesthetic perception, the human being does not only look at the world. It looks and experiences itself in the world, and experiences the world in itself. Presentative spatial experience does not permit distance. The aesthetic space of architecture is a space in which the human being experiences the density and breadth on his own body. The space of sound is only experienced as space within the body corresponding and resonating with the rising and falling, the swelling and fading of succeeding tones.

But how does the 'insularity' of aesthetic perception harmonize with our thesis about the underlying relational character of all consciousnesses? For Cassirer, there is no perception not caught up in a relation to others. Perception – and indeed consciousness – is always a linking. This 'entanglement' of perception is, in the case of aesthetic perception, an imprisonment. It is a moment of self-perception. For Martin Seel, it is the matter of the 'subjects of aesthetic perception and the craving of the real present through the interrogation of another's present. In the sensory presence of an object we notice a moment of our own present.'[66] Also, for Hans Ulrich Gumbrecht, what we feel in the intensive moment of aesthetic perception is 'probably nothing else but a specialized higher degree of the functionality of our general cognitive, emotional, and perhaps even psychic faculty'.[67] Who would not now think of Kant (who in his aesthetics emphasizes the interaction of many faculties rather than the functioning of one faculty)?

Unlike Kant, who is interested in the purview and borders of human reason, Cassirer is interested in representational and expressive forms of reason. He must therefore focus his sights on 'the mode of objectivity of the aesthetic object', its symbolic form.[68] From this perspective, art as a phenomenon builds, among other things, a universe of discourse, an integral of experience and a form of representation. Kant limited himself to the phenomenon of aesthetic perception because he wanted to demonstrate the rationality of aesthetic judgement.

Cassirer takes a two-pronged approach to adequately describe the complexity of the art-aesthetic phenomenon. On the one hand, Cassirer examines the phenomenon in the context of the symbolic

forms. In so doing, however, he does not neglect the irreducible unique-
ness of every aesthetic perception. He affirms that, '[a]ll other things are
lost to a mind thus enthralled; all bridges between the concrete datum
and the systematized totality of experienced are broken'.[69] But the aesthetic perception is still not without a relation. The connection through
which the perception is registered by consciousness depends on the
inner subjective relational character of subjective experience. Cassirer calls
this aesthetic state a special 'resonance' of the 'totality of the powers
of mind'.[70] The art-aesthetic phenomenon 'in one stroke presents that
unity of mood which is for us the unmediated expression of the unity of
our ego, of our concrete feeling of life and self'.[71] This feeling, though, is
only the fragile and passing moment of presence. What we have, then,
is the intersection of two orders: the inter-subjective extensive of symbolic form and the intensive monadic nature of self-consciousness.

Translated by Wilson McClelland Dunlavey

Notes

1. Michel Foucault, *Die Ordnung des Diskurses* (Frankfurt am Main: Fischer, 1991), 45.
2. The following remarks are based on my book *Ästhetik im Zeichen des Menschen: Die ästhetische Vorgeschichte der Symbolphilosophie Ernst Cassirers und die symbolische Form der Kunst* (Hamburg: Meiner, 2007).
3. Immanuel Kant, *Critique of Pure Reason*, B 303.
4. Ernst Cassirer, *The Philosophy of Symbolic Forms*, vol. 1: *Language*, trans. Ralph Manheim (New Haven: Yale University Press, 1971), 76.
5. Cassirer, *Philosophy of Symbolic Forms*, vol. 1, 7.
6. Ernst Cassirer, *The Philosophy of Symbolic Forms*, vol. 2, *Mythical Thought*, trans. Ralph Manheim (New Haven: Yale University Press, 1971), 12.
7. *Ibid.*, 4.
8. Ernst Cassirer, *Leibniz' System in seinen wissenschaftlichen Grundlagen*, in: Cassirer, *Gesammelte Werke*, ed. Birgit Recki (Hamburg: Meiner, 1998–2009), vol. 1, 418. For a comparison of Cassirer's philosophy of the symbol with leib-nizian Monadology, see the chapter 'Symbol und Ausdruck: Die Leibnizschen Quellen der Philosophie der symbolischen Formen', in Massimo Ferrari, *Ernst Cassirer: Stationen einer philosophischen Biographie*, trans. Marion Lauschke (Hamburg: Meiner, 2003), 163–82. For a general overview, see Martina Plümacher, *Wahrnehmung, Repräsentation und Wissen: Edmund Husserls und Ernst Cassirers Analysen zur Struktur des Bewußtseins* (Berlin: Parerga, 2004).
9. See Cassirer, *Leibniz' System*, 416f, and Cassirer, *Philosophy of Symbolic Forms*, vol. 1, 33.
10. Cassirer, *Leibniz' System*, 417. The quote is translated into English by Wilson McClelland Dunlavey. Subsequent quotes that are translated by Dunlavey will be marked 'trans. W. Dunlavey'.

11. Ernst Cassirer, *Substance and Function*, trans. William Curtis Swabey and Marie Collins Swabey (New Haven: Yale University Press, 1953), 281.
12. Cassirer, *Substance and Function*, 284.
13. Cassirer, *Philosophy of Symbolic Forms*, vol. 1, 100. Plümacher therefore speaks of a 'semiotically oriented epistemological internalism' in Cassirer, see Plümacher, *Wahrnehmung, Repräsentation und Wissen*, 485.
14. Cassirer, *Substance and Function*, 284.
15. See Johann Wolfgang von Goethe, 'Vermächtniß', in *Werke* (Weimar: H. Böhlau, 1887–1919), part. I, vol. III, 82f, and in Goethe's letter to Karl Friedrich Zelter from 31th December 1829, in *Briefe*, part IV, vol. XLVI, 197–200, 199: 'I have realized that I hold fruitful thoughts – those that attach themselves to my thoughts and, at the same time, further them – as true thoughts' (trans. W. Dunlavey).
16. Cassirer, *Substance and Function*, 284.
17. *Ibid.* In recent research in the philosophy of science, the concept of representation is used similarly. A representation is here not understood as a presentation 'of something' but a presentation 'in something'. See Hans-Jörg Rheinberger, Michael Hagner, Bettina Wahrig-Schmidt (eds), *Räume des Wissens: Repräsentation, Codierung, Spur* (Berlin: Akademie Verlag, 1997), 9f: 'The molecular turn in biology since the middle of the 20th century appears to imply that the traditional relation between representation and reference might have been turned upside down insofar as the molecular script can no longer be thought of as a presentation of something. Instead it is the primordial instance that first generates the representation … Representation realizes itself in completely different forms than experimental order.'
18. See also Andreas Graeser, *Ernst Cassirer* (München: Beck, 1994), 31.
19. Cassirer, *Philosophy of Symbolic Forms*, vol. 1, 97.
20. *Ibid.*, 98.
21. *Ibid.*
22. *Ibid.*, 105.
23. Ernst Cassirer, *Philosophy of Symbolic Forms*, vol. 3: *The Phenomenology of Knowledge*, trans. Ralph Manheim (New Haven: Yale University Press, 1957), 128–9.
24. In this connection, see also Oswald Schwemmer, *Ernst Cassirer: Ein Philosoph der europäischen Moderne* (Berlin: Akademie Verlag, 1997), 89–107.
25. Cassirer, *Philosophy of Symbolic Forms*, 1, 107. In *Zur Logik des Symbolbegriffs*, from 1938, Cassirer answers the critique of Konrad Marc-Wogaus by clarifying that the separation between presentation and representation is a 'relative separation' or an 'ideal separation'. See Ernst Cassirer, *Gesammelte Werke*, ed. Birgit Recki (Hamburg: Meiner, 1998–2009), vol. 22, 120f. Elsewhere Cassirer makes it clear that it is impossible to uncover a layer of immediate experience. Present and representative only display themselves in their attachment. See Ernst Cassirer, *Nachgelassene Manuskripte und Texte*, eds John Michael Krois and Oswald Schwemmer (Hamburg: Meiner, 1995–2009), vol. 1, 48f. In fact, to differentiate a sensory consciousness from a mental one is problematic. Oswald Schwemmer interprets the relation between presentation and representation as a relation between 'primary' ideational realization and secondary realization, in Schwemmer, *Ernst Cassirer*, 95ff. The formulization of Cassirer's that he quotes appears

to justify this interpretation. A process in which the artistic symbolic 'interprets' the natural symbolic is therefore hypothetical. A 'primary' ideational realization cannot actually be 'conscious' – without negating the stated relational character of consciousness. Such a kind of sensory consciousness must be rejected. See also Dominic Kaegi, 'Jenseits der symbolischen Formen: Zum Verhältnis von Anschauung und künstlicher Symbolik bei Ernst Cassirer', *Dialektik* 1 (1995), 73–84.

26. Cassirer, *Philosophy of Symbolic Forms*, vol. 3, 90.
27. Cassirer, *Philosophy of Symbolic Forms*, vol. 1, 29 and 41.
28. *Ibid.*, 42, and *Philosophy of Symbolic Forms*, vol. 3, 126.
29. Cassirer, *Philosophy of Symbolic Forms*, vol. 3, 115.
30. *Ibid.*, 81.
31. *Ibid.*, 62 (and 65).
32. *Ibid.*, 68.
33. *Ibid.*, 3.
34. Cassirer, *Philosophy of Symbolic Forms*, vol. 1, 111.
35. Cassirer, *Philosophy of Symbolic Forms*, vol. 3, 115.
36. It makes no difference whether we are talking about a rain dance, stabbing a voodoo doll or the Catholic Eucharist.
37. Cassirer, *Philosophy of Symbolic Forms*, vol. 3, 112.
38. *Ibid.*, 110.
39. *Ibid.*, 307.
40. For a convincing defence of 'reference' in literary criticism, see Eckhard Lobsien, 'Mimesis und Referenz: Paradigma Ulysses', *Zeitschrift für Ästhetik und Allgemeine Kunstwissenschaft*, 50:2 (2005), 227–43. Lobsien understands reference – or what is represented in literature – to be something that 'is realized in the presence of the circumstance referred to', 242. All representations – artistic ones included – are for him (and Cassirer) 'primary definitions': 'Representation produces that which is represented. The representation is necessary for the evocation of the represented' (235f). He develops a minimal definition for the concept of 'reference' that is only essentially fulfilled in the use of a non-simulated name. Meaning accrues first through the linking and embedding of this name in connection with a formulation, by which it is simultaneously transcended (238). 'Reference is a singular empirical fact. The expression realized by this reference bears this fact in a context that is not a reference, but rather an elementary model of the world' (241).
41. Jacques Derrida, *Of Grammatology*, trans. Gayatri Chakravorty Spivak (Baltimore: John Hopkins University Press, 1997), 7.
42. Uwe Spörl, 'Konzeptionen von Performanz im Rück- und Ausblick', *KulturPoetik*, 5:1 (2005), 95–7 (trans. W. Dunlavey).
43. Dieter Schlenstedt, 'Darstellung', in: *Ästhetische Grundbegriffe*, vol. 1 (Stuttgart: J. B. Metzler, 2000), 838 (trans. W. Dunlavey). Schlenstedt here quotes Samson S. Sauerbier, *Gegen Darstellung – Ästhetische Handlungen und Demonstrationen: Die zur Schau gestellte Wirklichkeit in den zeitgenössischen Künsten* (Köln: Walter König, 1976), 247 (trans. W. Dunlavey).
44. Karl Heinz Bohrer, 'Die "Antizipation" beim literarischen Werteurteil: Über die analytische Illusion', in Bohrer: *Plötzlichkeit: Zum Augenblick des ästhetischen Scheins* (Frankfurt am Main: Suhrkamp, 1981), 31 (trans. W. Dunlavey).

45. Hans Ulrich Gumbrecht, *Production of Presence: What Meaning Cannot Convey* (Stanford: Stanford University Press, 2004), 12.
46. *Ibid.*, 53.
47. See Hans Ulrich Gumbrecht, *Diesseits der Hermeneutik: Die Produktion von Präsenz* (Frankfurt am Main: Suhrkamp, 2004), 12. He discusses the terminus of Heidegger's notion of being in the world, in order to explain the void of distance in aesthetic perception.
48. See Gernot Grube, *Repräsentationen: Skizze für einen relationalen Repräsentationsbegriff unter kritischer Bezugnahme auf Ernst Cassirer und Nelson Goodman* (Berlin: dissertation.de, 2002), 56f: 'The concept of representation is at the center of Cassirer's philosophy of symbolic forms. To be precise, one must say that Cassirer builds a theoretical center, because this decisive conceptual instrument strikes throughout the whole world.'
49. Enno Rudolph, 'Metapher oder Symbol – Zum Streit um die schönste Form der Wirklichkeit: Anmerkungen zu einem möglichen Dialog zwischen Hans Blumenberg und Ernst Cassirer', in Reinhold Bernhardt, Dietrich Ritschl and Ulrike Link-Wieczorek (eds), *Metapher und Wirklichkeit: Die Logik der Bildhaftigkeit im Reden von Gott, Mensch und Natur* (Göttingen: Vandenhoeck und Ruprecht, 1999), 326 (trans. W. Dunlavey).
50. Cassirer, *Substance and Function*, 284.
51. Paul Celan, 'Ansprache anläßlich der Entgegennahme des Literaturpreises der Freien und Hansestadt Bremen', in *Gesammelte Werke in sieben Bänden* (Frankfurt am Main: Suhrkamp, 2000), vol. 3, 186 (trans. W. Dunlavey).
52. Susanne K. Langer, 'De profundis', *Revue International de Philosophie*, 110:4 (1974), 439.
53. Christiane Schmitz-Rigal, 'Modi des Symbolischen und plurale Sinnwelten: Zum Verhältnis der symbolischen Formen Ernst Cassirers', *Allgemeine Zeitschrift für Philosophie* 29:3 (2004), 249–61, 251 (trans. W. Dunlavey).
54. Cassirer, *Philosophy of Symbolic Forms*, vol. 3, 62 and 65.
55. *Ibid.*, 68.
56. *Ibid.*, 264. Cassirer makes this decision in the context of his investigation of the pathology of symbolic consciousness. I mean to suggest that in other places Cassirer suggests that the 'presentative mental attitude' need not exclusively be judged as pathological.
57. Ernst Cassirer, *Die Philosophie der Aufklärung*, in *Gesammelte Werke*, vol. 15, 315.
58. Cassirer, *Philosophy of Symbolic Forms*, vol. 3, 3.
59. Cassirer, *Philosophy of Symbolic Forms*, vol. 2, 35–6.
60. Cassirer, *Philosophy of Symbolic Forms*, vol. 3, 68.
61. See 'Language and Art I', in Donald Phillip Verene (ed.), *Symbol, Myth, and Culture: Essays and Lectures of Ernst Cassirer, 1935–1945* (New Haven: Yale University Press, 1979), 157. See also 'Language and Art II', in Verene, *Symbol, Myth, and Culture*, 186.
62. Cassirer, *Philosophy of Symbolic Forms*, vol. 3, 94.
63. Cassirer, *Philosophy of Symbolic Forms*, vol. 2, 43 (italics mine).
64. Karl Heinz Bohrer, 'Vorwort', in Bohrer, *Plötzlichkeit: Zum Augenblick des ästhetischen Scheins* (Frankfurt am Main: Suhrkamp, 1981), 7.
65. Cassirer, *Philosophy of Symbolic Forms*, vol. 3, 171.
66. Gumbrecht, *Diesseits der Hermeneutik*, 119.

67. Martin Seel, *Ästhetik des Erscheinens* (Frankfurt am Main: Suhrkamp, 2003), 62.
68. Ernst Cassirer, *Kant's Life and Thought*, trans. James Haden (New Haven: Yale University Press), 310.
69. Ernst Cassirer, *Language and Myth*, trans. Susanne K. Langer (New York: Dover Publications, 1953), 58.
70. Cassirer, *Kant's Life and Thought*, 317.
71. *Ibid.*, 316.

9
Cultural Poetics and the Politics of Literature

Frederik Tygstrup and Isak Winkel Holm

In his memoirs, *A Tale of Love and Darkness*, published in 2004, Amos Oz tells the story of his grandmother's death. Arriving in Israel on a warm summer's day in 1933 from one of Eastern Europe's grey winter villages, she saw the hot marketplace in front of her, with its bloody carcasses, colourful fruit, sweating men and noisy vendors and passed her verdict: 'The Levant is full of microbes.' She immediately embarked on a comprehensive hygiene regime, which she zealously came to maintain over the next 50 years – a regime which included cleaning, scalding, airing and disinfecting everything, including her own body, on a daily basis. The cleaning frenzy comes to an end only when she collapses at 80-something with heart failure during one of the three hot baths, which were part of her daily routine. So what did the grandmother die of? The fact is that she died of heart failure. But the truth is that it was her monstrous hygienic programme that killed her. And, on a philosophical tone, Oz adds: 'Facts tend to hide the truth from our eyes.'[1]

Truths and facts

The world is full of facts, and we are presented with still more facts at a still faster pace. For the last couple of centuries, facts have been the undisputed starting point of any knowledge. We consider facts as 'hard' and see them as a solid foundation for our ideas and actions. But the encounter with the evidence of the fact is followed by the questions of *how* and *why* this fact comes about – questions to which the fact in itself does not necessarily provide us with an answer. Facts must be organized into patterns for us to relate to them. In order to understand the grandmother's heart failure we must know about the manic cleanliness which led to the hot baths and thus put a fatal strain on her circulatory system. And in order to understand her manic cleanliness, we must take into account a complex set of conceptions about the East and the West, about corporeality and sensuousness, about visibility and invisibility, about control and vulnerability and so on. The truth Amos Oz is seeking does not immediately follow from the facts, it only begins to emerge when a host of such diverse secondary circumstances – material, psychological, historical or cultural factors – are combined with the facts in that specific configuration which for him is able to capture the grandmother's story.

The facts in place, the true image of reality is still to be produced. This production of images of reality can be considered both on a small and on a large scale. On the small scale it could be a matter of an author's configuration of facts into a coherent shape by telling stories or setting up complex metaphorical patterns. According to Oz, writing a novel is like building a full model of a city from matches – complete with buildings, squares, boulevards, towers and suburbs, down to the smallest bench.[2] And it is the consistency of the model that enables us to understand the individual element. Today, when facts are accessible to us on an unprecedented scale, Oz suggests, our need is greater than ever for fictions and narratives that can provide the facts with a world in which they make sense.

On a larger scale, a culture similarly produces collective images of reality by capturing facts and integrating them into interpretive frameworks, quite like the work of the individual author. Any culture possesses a common set of models prescribing our ways of seeing, of thinking and of relating to ourselves, to each other and to the shared surroundings. Living in a culture very much amounts to having acquired this common repertoire of narratives and cognitive forms that can be used to configure facts. These collective techniques and principles enabling the production of cultural images of reality may be called, with Louis Montrose's felicitous term, a 'cultural poetics'.[3]

We never create our individual frameworks of understanding in a cultural vacuum. In each case, a cultural poetics will be at hand with a series of suggestions for already existing ways in which to frame a fact. An individual fact may be spectacular, unprecedented or peculiar; but the mere inertia of the familiar cultural repertoire of accepted truths and explanatory models will more often than not reassure us that the fact can be identified with something we already know. In such cases the situation is virtually the opposite of the one diagnosed by Amos Oz: here it is no longer the facts that obscure the truth, but rather the already established truths which conceal or mask the singularity of a given fact.

In this sense, cultural poetics not only offers to arrange the facts and render them comprehensible in a neutral manner; cultural poetics *produces* facts. It indicates and determines what is a fact and what is not, thereby constituting a cognitive window on reality through which certain facts are accessible and others remain outside the field of vision. In the light of a given cultural regime of truth, facts may be meaningless, inconceivable, frivolous or heretical; consequently, they will remain in the dark. The cultural poetics constitutes a system of inclusion that captures facts and gives them an explanation and a framework, but it also constitutes a system of exclusion that determines what can rightfully be perceived as a fact, and what should be excluded from this category. Michel Foucault highlighted this historical mechanism of exclusion in his comments on Mendel's theory of heredity; Mendel introduced a concept of hereditary features which did not correspond to any existing scientific categories, and whose function broke with contemporary conceptions about reproduction. What Mendel said was true, Foucault remarks, but it was not 'within the truth': that is, not in accordance with the model for biological scientific truth subscribed to by his contemporaries.[4]

If an author wants to give a true image of reality, then, siding with the truth against the contemporary avalanche of facts is not enough. In order to say something true, one may also have to break out of the truth of one's contemporaries. When Salman Rushdie, in *The Satanic Verses*, refers to a city which is 'visible, but unseen',[5] he sides with the facts against the received truths. Rushdie's implication is that in a modern metropolis, there are so many facts and so many complicated relations between them about which we know very little, because we prefer to decipher and understand the world in the same way as generations before us. Even if this wonderfully complex urban reality is visible to anyone, it nonetheless remains unseen, simply because we do not possess an adequate way of viewing it. Our cultural poetics has not trained us sufficiently to actually see what is visible. Here the facts challenge the truth, and the task of the author is, thus, to highlight facts and to insist on focusing on them until it actually becomes possible to see them.

Amos Oz defends truth against facts; Salman Rushdie defends facts against truth. Nonetheless, the two authors have a common concern, namely an interest for how literature may create an effective image of reality in the continuous friction between truths and facts, between cultural knowledge and practical experience. In this, the two novelists pose the classic question of representation. Literary studies have traditionally discussed this question using concepts such as mimesis, realism and reflection. Representation is not only the concern of literary scholars, however, but exists in the two formats described above, as a literary poetics and a cultural poetics. Literary poetics concerns the techniques used to represent reality within the relatively narrow confines of literature. In contrast, cultural poetics describes the pervasive principles for interpreting reality which apply across the different spheres and institutions of a society. Whereas the literary poetics is typically characterized by a sharply focused individual intention, the cultural poetics is, rather, an anonymous collective work without a distinct sender. Seen from the perspective of representation theory, however, the two poetics work in much the same manner: both are engaged in creating forms that can make a changing reality accessible to thought and action.

In the following, we will describe the position of literature in relation to 'the political' by discussing the interaction between literary poetics and cultural poetics. First, we will characterize the cultural poetics more precisely through the concept of 'symbolic forms'. Second, we will focus on the literary poetics and its institutional context. And finally, in a third step, we will take a closer look at the interaction between literary and cultural poetics.

Symbolic forms

In philosophy, the discussion about truths and facts is often identified with the quarrels between rationalist and empiricist thinkers in the eighteenth century about the foundation for knowledge, either in the indisputable truths of reason, as rationalists have it, or in the facts of empirical observation. And it is in Kant's twofold critique of his predecessors that we find the first contours of the problem of representation outlined above. To Kant, the essential is neither truths nor facts, but rather the laborious production of images of reality that takes place in the space between them. According to Kant, cognitive experience is a process configuring the material of sensory perception in such a way that it forms a meaningful unity. And this process follows specific procedures. The work of combining the categories of reason with the phenomena of intuition is regulated by a series of recipes for how the manifold in the sensuous intuition can be assembled into a unity. Using a modern metaphor, we may say that, according to Kant, human consciousness is equipped with its own unique operating system that sets the conditions for how categories and phenomena can be made coherent.

In *Critique of Pure Reason,* Kant explains that the transcendental operating system consists not only of categories, but also of figures. Cognition relies on a type of pure, abstract figures, which Kant calls schemata. The schema is a figment of the productive imagination enabling a conceptual grip of the manifold of the sensuous intuition. It is a 'monogram', a figurative matrix for the combination of the sensuous diversity issuing from perception. In Kant's words, our use of such transcendental forms 'is a hidden art in the depths of the human soul, whose true operations we can divine from nature and lay unveiled before our eyes only with difficulty'.[6]

Unfortunately, Kant does not elaborate on the schematism of reason. The mapping of this hidden art was only performed 140 years after the publication of *Critique of Pure Reason,* by Ernst Cassirer in *The Philosophy of Symbolic Forms.* Like Kant, Cassirer argues that a workable image of reality should be based neither on the concepts of reason or the data of perception, but should be built in the field between them. However, according to Cassirer, schematism should not be understood in a purely logical manner, as Kant would have it; it must be understood historically, on the basis of the actual historical images of reality that can be uncovered by the study of cultural history. While thus historicizing the Kantian critique of knowledge, Cassirer at the same time reverses the Kantian conceptual hierarchy. For Kant, the starting point was the rational concepts and the sensual perceptions that had to be fused in the magic of intuition. For Cassirer, things work the other way round: his starting point is the existence of human images of reality, which are used as a basis for reconstructing certain conceptual systems and sensory forms of attention.

Cassirer's term for such images of reality is 'symbolic forms'. Symbolic forms constitute a historical schematism that operates as the transcendental condition of possibility for the interpretation of reality in a specific culture. 'Cultural poetics' can thus be understood as a specific historical regime of symbolic forms. Symbolic forms is a cultural repertoire of mental models allowing us to assess reality and thus to deal with it pragmatically in everyday life. Symbolic forms differ widely in terms of their scale and scope; as shorthand examples, we can consider the confession as a symbolic form which makes it possible to form an image of a person's intimate life; the diagnosis as a symbolic form which regulates the understanding of the relationship between the normal and the pathological; or the social contract as a formula which sets up rules for human relationships.

To engage in an analysis of symbolic forms, Cassirer takes the actual historical cultural practice as his starting point. He explores the concrete ways in which reality, in specific places and at specific times, has been described by those experiencing it. On this point, Cassirer shares a conviction with the German *Geistesgeschichte* of his time: that a given culture should be understood from within, in the light of its actual values, insights and practices, rather than from the outside, through some general principles of description. At the same time, though, the cultural history of symbolic forms also differs from *Geistesgeschichte* by way of its materialism. For Cassirer, the ambition is not to reconstruct a mental landscape of ideas and intuitions, but to piece together an actual textual and visual archive of historical images of reality. In other words, Cassirer's subject matter is not *ideas* about reality, but *representations* of reality. For instance, the important thing is not the confession as an underlying mental template, but, rather, the plethora of actual confessions in their available textual manifestations.

The philosophy of symbolic forms also displays a striking similarity with another feature of *Geistesgeschichte*, namely a strong tendency to homogenize the image of a culture by focusing on cohesion and totality rather than on contrasts and confrontation. As a corrective, one might reiterate Marx's point – put forward in *The Eighteenth Brumaire of Louis Bonaparte* – that an image of reality does not necessarily include all the different potential perspectives on this reality. *Darstellung* is not the same as *Vertretung*; representations of reality are not necessarily representative of the represented reality.[7] In other words, the validity and impact of a cultural system of symbolic forms will remain determined by a social order of power and social interests. Power relations are involved in the favouring of certain ways of thinking, speaking and acting; and in the exclusion of others as meaningless or trivial. A repertoire of symbolic forms offers the opportunity to make a range of different utterances, but not all of these will be perceived as meaningful and representative statements about the order of things. Seen in this light, the social struggle between different interests and social groups can be described as a struggle between ways of forging images of what reality looks like and what it ought to look like. This conflictual aspect is attenuated in Cassirer's description of rather homogeneous cultural totalities, whereas Italian political philosopher Antonio Gramsci's concept of hegemony helps keep it in focus. Symbolic forms are hegemonic when their way of representing reality serves the interests of the dominant powers.[8]

The political dimension of symbolic representations of reality is not only visible in this massive, agonistic sense, however, where class interests are lined up in confrontation to seek hegemony for their own interpretation of the world. The poetics of culture is played out in the public arena where political ideologies, works of art and scientific hypotheses are presented; it also lies at the basis of the individual's everyday life at work and in private contexts. The specific models for how the things of the world can be configured also determine how objects and people are categorized in the intimate and private spheres of human life, as well as in society at large. In this sense, the political is omnipresent and by no means limited to the 'political' topics discussed in newspapers and in parliamentary assemblies. This insight can be captured by Jacques Rancière's concept of the division of the sensible. Rancière's 'partage du sensible' is a name for the basic order of things, the 'a priori forms' – as he puts it with specific reference to Kant – 'which determine what presents itself to sensory perception and defines the coordinates for time and space, as well as for what can be said and what can be seen'.[9]

Finally, the poetics of culture is not only a question of rational knowledge, but also of desire. The repertoire of symbolic forms is not only a toolbox of antiseptic instruments designed to deal with the world in an objective manner; it is also a billowing veil of fascinations, dreams, aversions and idealizations. A cultural poetics can produce wonderfully meaningful and coherent images of reality by projections of desire and paranoid illusions. Desire distorts the cultural filter of modes of imagination and ways of thinking, which means that the images of reality contain patches of both inexplicable blindness and hysterical sharpness. This insight is highlighted in Slavoj Žižek's concept of unconscious ideological fantasy. In Žižek's Lacanian optics the collective fantasies function as the condition of possibility for the ability to perceive, act within and talk about the surrounding society. He talks about the sociosynthetic function of fantasy, and the formulation has clear roots in Kant's theory of cognition. The same goes for Žižek's characterisation of these fantasies as a formal-transcendental *a priori*.[10]

The symbolic forms constitute the hidden art – not in the depths of the human soul, as Kant put it, but in the depths of human culture. This hidden art is a vital part of the political. In the common sense of the term, the political covers the familiar game of views and values, of conflicts and connections. In a wider sense, however, it also includes the images of reality that determine what social issues and what social conflicts are perceived as pressing reality. Cultural poetics is a political poetics, insofar as it creates specific images of social reality and in the same process elides other facets of reality. In other words, politics is not just the art of the possible, it is also the hidden art which makes it possible to perceive of the real in the first place.

The politics of literature

The poetics of literature is simultaneously part of and apart from the poetics of culture. Literary poetics is a part of cultural poetics, because it contributes to the production of images of reality that can serve as models for other claims to an understanding of the world. On this level, there is no essential difference between literature and other symbolic forms. To be sure, one can easily distinguish between a tragedy in blank verse and the minutes from a board meeting, or between a sonnet and a feature article, but this is due to the existence of traditionalized genre markers rather than to any ontological features pertaining specifically to literature. Fundamentally, it is really the same – and remarkably limited – repertoire of cognitive models that functions across the various cultural spheres and institutions. For example, the confession has played an important role in the literary autobiography since Augustine, but the same symbolic form is also found in the confessional, in the interrogation room, in psychotherapy, in the television show, in the conversations struck up in a pub.

At the same time, though, literary poetics is not only part of, but also apart from the cultural poetics. When it is possible to distinguish quite easily between literary and non-literary utterances, it is because individual linguistic phenomena not only have a form, but also a framework. This framework is an institutional regulation of the social function of different utterances in respect to how they are produced, how they circulate and how they are used. The institutional framework puts the symbolic forms to work by using them as tools in a given social web of power relations and political interests. In the Catholic Church the confession serves a strictly defined purpose; in the modern legal system the confession has a different, although no less specific function.

Literature is distinctive due to its framework, not to its form. From a sociological point of view, literature is an institutionalized social field with its own rules, its own economy and a particular societal existence. The literary institution is a conglomerate of social routines, including the production, distribution and consumption of literature, and a vital part of this institution is a collective agreement to regard certain texts with a specific attitude of mind. This agreement is normally confirmed by endowing these texts with certain material, so-called 'paraliterary' markers, through which they can be recognized. The institutionally empowered convention gives these texts the status of 'fiction' – that is, texts whose pragmatic and referential functions differ from other texts.

Jacques Derrida has captured this problem by describing literature as both a fictional institution and an institutionalized fiction.[11]

The institutionalization of fiction establishes a crucial distance to the surrounding cultural poetics by suspending any unambiguously pragmatic function; utterances that circulate within the literary institution have no clearly defined social significance and impact. This institutional distance enables literature to adopt a non-pragmatic and non-intentional relationship to the reality it depicts. The literary poetics may draw on the same symbolic forms as the surrounding cultural poetics, but when these forms appear in literature, they appear suspended and without any clearly assignable accountability. Rousseau, Dostoyevsky, Camus and Coetzee do not make use of the confession for the purpose of having a particular individual judged or initiating a specific therapeutic process.

On the level of symbolic forms, to sum up, literary poetics is part of a cultural poetics, but on the level of the institutional framework, the two poetics part company. Literary images of reality generally make use of the same symbolic forms as other cultural interpretations of reality, but in a different manner. Thanks to this duality of likeness and difference, literature takes the guise of a cultural laboratory where experiments can be conducted, testing, as it were, different roles and functions of a cultural repertoire of images of reality. Paradoxically, literature's irresponsible distance to society is a condition for its ability to intervene in the life of society, not necessarily by voicing specific political views, but rather by reflecting, varying and contesting the dominant cultural poetics' division of the sensible.

The same line of thought can be expressed by saying that literature *deviates* from the way in which cultural images of reality are generally produced. In twentieth-century literary theory, the concept of deviation – described as alienation, defamiliarization, transgression, estrangement and so forth – has been among the most important bids for identifying the so-called *literariness* that sets literary language apart. According to this theory, literary images of reality differ from the more automatised representations of reality in society at large. Theories of deviation usually deal with stylistic parallelisms and metaphorical paradoxes, but literary deviation takes place not only on the linguistic level; deviation also operates on a more fundamental level that concerns the cognitive schemata through which an image of reality is produced. Literature can transform not only the stylistic form, but also the underlying symbolic form of a representational routine. Due to its institutionally sanctioned distance to the pragmatics of social life literature can change our everyday and habitual – and therefore hidden – ways of constructing images of reality.

Creation, exposition, transposition

On a general note, we will propose to distinguish between three different modes of deviation on the level of symbolic forms (modes which can, of course, overlap and appear side by side in a literary work). First, deviation can take the shape of a *creation* of new images of reality that challenge a dominant regime of symbolic forms. Since the material of literary representation of reality is always a historical language, the creation of new models is always an adaptation of existing models, a constructive reorganizing of pre-existing material. Here, deviation operates a de-formation and trans-formation of prevalent ways of constructing images of reality. Fiction here serves to think and show something that does not naturally fall within the habitually conceived order of things. This attitude was the starting point for both Amos Oz and Salman Rushdie, circling about something they want to express, as it were – how the grandmother's life is reflected in her death, how different mental universes converge in creating a specific spatial quality – but which cannot be grasped with the available modes of explanation and description. In order to approach aspects of reality that would otherwise have remained silent, they must write new stories, invent new forms of description. Giving a voice to a truth about a singular life or a singular event requires that an aspect of reality be liberated from the standard cultural repertoire of themes and forms. What is required, in other words, is the creation of a specific and unheard-of configuration of events, emotions and viewpoints capable of grasping the way in which the singular deviates from the general.

Literature creates experimental models of reality. These experiments originate in the singular: the concrete configuration of forms, of lines, colours, shapes, sounds or words. Oz's image of the grandmother springs from a single statement: 'The Levant is full of microbes'; Rushdie's image of the city starts with an encounter on a street corner. The singular artistic starting point opens up a potential window on the world. By composing an image of the world from this viewpoint, the work of art is able to not only describe and account for the individual situation – how the grandmother's life converges and makes sense in the light of the simple statement – but also to create a far more comprehensive explication of reality. To the extent that an image of the 'visible, but unseen' reality is successfully created, an entire image of a world emerges. How do Israeli everyday life, political life and mental universes look in the light of the grandmother's life story, as told by Oz? Or, how can we observe the postcolonial city as a conglomerate of material, mental and social layers?

Literature's capacity to create new models of reality enables it to alert its readers to hitherto unseen aspects of social reality. Literature has rendered visible the reading and writing daughters of the bourgeoisie in the eighteenth century, the entrepreneurial third class in the nineteenth century, industrial mass societies in the first half of the twentieth century, and the postcolonial migrants from the 1950s until today. The artwork's division of the sensible begins in the configuration of the single image, of the single perceptual starting point, but the division has a potential and wide-reaching resonance in the work, in the reader, in tradition and in social life.

Second, the deviation can take the shape of an *exposition* of a culture's repertoire of images of reality. From this perspective, literature is not only a laboratory for the creation and transformation of symbolic forms, it is also a medium capable of rendering existing forms visible. The exposition can be regarded as a sort of citation of existing images of reality, and as such it can take many different forms, from direct citations of historical uses of language to reconstructions of 'typical' images of reality. In his works on realism, Georg Lukács formulated a concept of 'the typical', inspired by Max Weber's sociological concept of the ideal type. To Weber, the typical did not necessarily constitute something that is typical in a statistical sense; rather, it summarizes some essential features that are not likely to be matched by any empirical findings. For Lukács, the successful realistic character description was typical in this sense, as found, for example, in Balzac's works, which included the typical banker, the typical journalist, the typical countess and so forth. The typical characters of Balzac's novels gave a convincing impression of being real, because they were provided with an abundance of such individual oddities and peculiarities that are typical of the individual.[12]

A corresponding form of literary typicality is found on the level of symbolic forms, where the typical summarizes some essential features of a culture's way of understanding existential, social or material reality. Against this background, the technique of exposition can be described as an art of characterization by means of which the author does not modify, but rather condenses salient forms of a historically situated cultural poetics. This is the case when Flaubert cites the conventional language of the cattle show intertwined with the equally conventional language of the love encounter, when Hermann Broch, in *The Sleepwalkers*, cites the language of the public house, or when Chuck Palahniuk cites the language of sexual therapy. In these instances, the exposition of images of reality has replaced the functional pragmatics of the image of reality. When the symbolic forms of a culture are not only applied, but also exposed, transformed into typifying quotations, the invisible quotation marks around them become visible. The words and sentences appear as if surrounded by a fine luminous edge, like an ethnographical object in a display case. This is the alienating effect of the expository function, namely the making visible of the symbolic forms as what they are: flawed conventions upon which we depend heavily for our perception of reality. Most often, the hidden art is buried in the depths of human culture, but sometimes, literature is capable of unearthing this deeply hidden art, putting it on display and making it subject to debate.

Third, and finally, the literary deviation can take the shape of *transposition* of an image of reality from one institutional context to another. The transposition consists in moving a symbolic form across the boundary between the literary and the cultural poetics, thus negotiating the existing boundaries between literature and the social institutions surrounding it.

The most obvious example is the question of censorship. From the trials against Flaubert and Baudelaire to the fatwa against Salman Rushdie, a standard argument in the fight against censorship has been that the statements found in literature cannot be assessed in the same manner as the social and political statements circulating outside the literary institution. This strategy positions itself on the fine line separating between cultural and literary poetics. The image of reality – an unfaithful provincial wife or a boozing prophet – makes perfect sense on both sides of the boundary, but with dramatic differences. Is the image of Emma Bovary a celebration of a promiscuous provincial wife or a diagnosis of how wrong things can go for someone who is bored by provincial life (as Flaubert, somewhat hypocritically, claimed during the trial)? Is the image of the prophet an accusation directed at a historical person for indulging in dubious moral conduct, or does it present the impression of Muhammad of a delirious mind of the late twentieth century?

The transposition of an image of reality creates an effect of defamiliarization, merely through the change of context. One may try, for example, as Robert Musil once suggested, to quote a poem at a general assembly, or to place a crystal glass on a ploughed field.[13] Alienation effects of this type are not only something literature is more or less unfairly subjected to, as in cases of censorship; it is also something literature has been able to exploit. Literature can intervene directly in political affairs and discussions, it can be agitational and polemical, and it can articulate directly political positions – while enjoying the protection of its special status as literature. Thus, what literature says in relation to a political context remains unsaid in a certain sense, since it is protected by the attenuating suspension mechanisms of the literary institution. On the other hand, the system of attenuation has regularly annoyed authors eager to intervene because the subduing of resonance also means that the effect of any literary political intervention is relatively modest.

In our general overview we have outlined three forms of deviation, all of which exploit the boundary between literary and cultural poetics – and indeed the fact that literary poetics is simultaneously both a part of and apart from cultural poetics. The *creation* of fictional world models uses the privileged enclosure of the literary institution to deform and transform existing symbolic forms in the endeavour to think and show something that does not naturally fall within the given order of things. Here, literature is an eminently realistic undertaking in the sense that it tries to capture pieces of reality that haven't yet been named properly. The *exposition* of typical epistemic models and the exploration of their social function uses aesthetic distance to make evident the devices and social interests invested in the model in question. In this case, literature is realistic by reflecting and communicating existing historical images of reality. *Transposition*, finally, allows the literary models of reality to go beyond and question the institutional framework of literature. Here, literature is realistic insofar as it exceeds the boundary between fiction and reality and engages directly in political reality. The first two strategies of deviation – creation and exposition – depend on a reasonably stable boundary between literary and cultural poetics insofar as their thinking and showing of the world take place within an institution. In these cases, literature is political precisely by virtue of its status as literature. In contrast, in the third strategy of deviation, literature is political by exceeding this boundary and thus undermining its own status as literature.

The boundary dividing literary and cultural poetics has obviously been in constant flux since what we today agree to call literature was given a fairly well-defined institutional framework in the course of the eighteenth century. At times, literature has had great impact and been highly influential; at other times, it has remained socially rather insignificant. It has been tied to dominant ideological discourses, and to oppositional or suppressed ones; it has been closely connected to the consolidation of new historical identities, and it has withdrawn from all social interaction. In each of these cases, the actual exchange between literature as an institutionalized activity and the surrounding world has been renegotiated, and new boundaries have been drawn. Each new renegotiation can be described as a deviation questioning the functions, the forms and the possibilities of literary writing. Put differently, each work of literature relates, explicitly or implicitly, strategically or pragmatically, to its mode of existence as literature.

The boundary between literary and cultural poetics currently seems to be the subject of particularly lively negotiation. The fatwa against Salman Rushdie in 1989 may be taken as a symptom of this. In a Western context, the practising of religion has over the past couple of centuries taken place within a specific institutional context tied to the church and the private sphere, reasonably clearly separated from other contexts of social action. The fatwa suddenly introduced a religiously founded injunction, thus cutting right through the division between the religious, the social, the political and the literary, suspending all the differentiated transposition mechanisms.

This may indicate that the boundary between literary and cultural poetics is no longer upheld as something entirely robust, and that the current mixtures – not only of religion and politics, but also of politics and 'identity', of politics and advertising, of political interests and economic interests and, in a wider sense, of social production, consumption and reproduction – mark a new challenge in terms of the way literary practice should handle the boundary between literature and society at large. The documentary film genre is perhaps the art form that has most consistently exploited this transposition from 'fictional' images of reality to 'documentary' ones and vice versa. Similarly, literature also seems to challenge still more radically the boundary which defines it as literature – or 'mere literature'– and thus to be holding an altered literary practice up against the current destabilization or de-differentiation of the institutional boundaries of social life. Perhaps literature is becoming political in a new way: not as a question of the author's correct or incorrect views, but as a reaction to – and an exploitation of – the fact that the boundary between the literary and cultural poetics is undergoing rapid change: in other words, a politicization that not only stems from the practice of literary writing, but also from its mode of existence in society. This seems to have brought about a new form of involvement and a new creative impetus in literature, currently transporting images of reality and models for the construction of images of reality across the boundary between the two poetics. It is, not least, this development which prompts renewed reflection on the relationship between the poetics of culture and the poetics of literature.

Translated by Lise Utne

Notes

1. Amos Oz, *A Tale of Love and Darkness*, trans. Nicholas de Lange (Orlando: Harcourt, 2004), 45.
2. *Ibid.*, 331.
3. Louis Montrose, 'The Poetics and Politics of Culture', in *The New Historicism*, ed. H. Aram Veeser (London: Routledge, 1989), 15ff.
4. Michel Foucault, *L'Ordre du discours* (Paris: Gallimard, 1971), 37.
5. Salman Rushdie, *The Satanic Verses* (New York: Viking, 1988), 241.
6. Immanuel Kant, *Critique of Pure Reason*, trans. and ed. Paul Guyer (Cambridge: Cambridge University Press, 1998), B180.
7. See also Ernesto Laclau, 'Power and Representation', in Laclau (ed.), *Emancipations* (London: Verso, 1996), 84ff.
8. See Chantal Mouffe, *On The Political* (London: Routledge, 2005), 17ff.
9. Jacques Rancière, *Le partage du sensible* (Paris: La Fabrique, 2000), 13.
10. Slavoj Žižek, *The Sublime Object of Ideology* (London: Verso, 1989), 17 and 20.
11. Derek Attridge, 'This Strange Institution Called Literature: An Interview with Jacques Derrida', in *Acts of Literature* (New York: Routledge, 1992), 36.
12. Georg Lukács, 'Verlorene Illusionen', in *Balzac und der französische Realismus*, *Werke*, vol. 6 (Neuwied: Luchterhand, 1965), 478ff.
13. Robert Musil, *The Man Without Qualities*, trans. Eithne Wilkins and Ernst Kaiser (New York: Perigee, 1980), vol. 1, 231.

10
Cave Art as Symbolic Form

Mats Rosengren

Outside

In this text I will try to combine at least two, perhaps even three different interests. First, my curiosity as to how Cassirer's thought may prove its worth in helping to dissolve contemporary epistemological dead-ends; second, my long-time involvement with cave art studies and cave art; third, implicitly, my own doxological approach to contemporary theory of knowledge.[1] I will try to achieve this combination by briefly presenting what I see as two major problems within the study of cave art: first, the ubiquitous urge for an origin and, second, what I am going to call the mimetic curse in cave art studies; I will then proceed to show that Cassirer's ideas about symbolic forms and, more specifically, his philosophy of technology may help us to dissolve these problems, focusing mainly on his concept of 'organ-projection' and on the notion of symbolic forms understood as ever-changing ways to produce human reality; finally, I will comment upon the topicality of both Cassirer's thoughts and of cave art. My aim is to show that Cassirer's notions of technology and of symbolic forms may, to a certain extent, provide a new conceptual (and as it were doxological) framework for cave art studies, within which the centennial question of the sense of the traces found in the caves could be addressed anew in a productive way.

Entrance: the study of cave art

The Paleolithic cave art of northern Spain and southern France is among the oldest extant traces of mankind's symbolic endeavours. It is hardly surprising, then, that the images and the tracings, the engravings and the imprints found in the caves and shelters are also among the most difficult to grasp. The material at hand is both overwhelmingly rich and completely elusive, alien and yet all so familiar.

The Paleolithic cave art was discovered at least twice in modern times – first in 1879, when Sanz de Sautuola finally noticed the now famous bison in the ceiling of the cave of Altamira, a discovery that was first rejected as a forgery, then doubted for more than 20 years; and then again in 1901, when the notion of mural Paleolithic art was finally, but only after a series of decisive discoveries of painted caves, accepted as a scientifically respectable concept by the academic community. After this double, first material and then conceptual, discovery a multitude of different suggestions as to the why and the wherefore of all the different kinds of traces in the caves has been put forward. It began in 1901 with art for art's sake, and continued via sympathetic magic, totemistic demonstrations, sexualized mythologies, binary narratives up to – in the beginning of the twenty-first century – reports from shamanistic journeys and trance experiences.[2]

Strangely, and despite the lack of any kind of hermeneutical tradition bridging the time-gap,[3] the style of the images, the way they are dispersed in the caves as well as their purported realism seem to fit easily into our Western pictorial tradition, as established in the eighteenth century, consolidating the Paleolithic cave art of Spain and France as part of the vast symbolic form Art. This, however, is not the case with all prehistoric images. John Clegg, an Australian rock art researcher, describes the perceptual troubles involved in seeing Australian rock art:

Without knowing what to expect it is almost impossible to see what there is. A certain familiarity with styles of pictures is necessary to see them: a guide often has to point out form in great detail before a naive onlooker can see the kangaroo at all. Once the onlookers have their 'eye in', they know what to expect, and rapidly become expert at discovering items that have not been pointed out to them ... The processes of 'getting the eye in' involve familiarity with common natural markings on the rock, and understanding of the representational conventions used.[4]

This can, of course, be interpreted in at least two different ways: either you take Clegg's comment as an indication that there really, though still unbeknownst to us, is a cultural affinity, making European cave art more easily accessible for the average westerner – and start a quest for finding the 'missing links' back to the presumed 'cradle of art', be it Lascaux (approximately 17,000 before present (BP)) or the much older Grotte Chauvet (approximately 34,000 BP). Such a desire for an origin, and preferably a unique origin, has plagued cave art studies all through the twentieth century. Commenting upon its consequences, the philosopher Jean Louis Schefer writes: 'This phantasm of origin is grounded in a transfer of historic guilt, trying to reconcile or to think art without divorce (in a first paradise or in an initial tragedy) as a "natural", "instinctive" expression of the real.'[5] But, as Schefer points out a few pages earlier in his text, 'the primary function of the figures [that is, the pictures in the caves] is not to restore or to present the reality (to make a representation of it) – this point of view is that of an art-consumer of the 19th century (or of the dominating ideology of this art)'.[6] Accordingly, I think we should avoid even posing the barren question of when or where art was born. It leads us astray already through its implicit presumption – pointed out by Schefer – that such an origin would be possible to locate in time or space. But that is not all – this desire for a unique origin, combined with the presupposition that this original art is, in Schefer's words, a natural, instinctive expression of the real, excludes right from the start the possibility of seeing art (as well as all the other symbolic forms) as an always already present upsurge of meaning, as an always already ongoing creation of our human world.

Or you can focus upon the fact that *anyone* can get their eye in and adopt the stance of the cave art specialist Jean Clottes and make the following claim:

> Even persons who have never visited a painted cave and who have no interest in mural cave-art will immediately recognise the photography of a painted bison or of a chamois for what it is, that is a representation of an animal by a prehistoric artist, be that notion vague and imprecise.[7]

To explain the fact that pictures of cave art are readily seen *precisely as* pictures of cave art, Clottes correctly points out that, 'in our culture', all (or at least many) of us have seen images of cave art in books and journals, we have seen stamps with aurochs from Lascaux, different logos made from prehistoric motifs and so on. So far, so good. However, he continues with the claim:

> It is this kind of memories, a bit vague, that nourish our perception. They allow us to establish the category 'prehistoric art' where we instinctively range these figures. This is all the more remarkable

since, in this domain as in others, – for reasons of survival – Nature has not originally programmed us to perceive similarities but rather to notice unlikeness, to notice the exception rather than the norm. Thus, if we react differently when confronted with ice-age art, it is because this art presents a striking unity.[8]

Elsewhere, I have tried to untangle Clottes' multiple contradictions in the two passages just quoted.[9] Briefly put, Clottes' position is exemplary of what may be called the 'mimetic curse' in cave art studies – that is, the stubborn conviction that to see something is to passively reflect that which is already there, no additions or subtractions made, paired with the equally stubborn belief that mimetic representation always equals objective realism.[10] As is apparent in the texts just quoted, such a conviction may very well exist alongside a firm conviction that the categories of perception are culturally dependent. It is as if the cultural production of categories of perception – like the category 'prehistoric art'– is seen as the natural reflection of our 'natural programming', to use Clottes' phrasing. And then there is nothing surprising in the fact that Paleolithic cave art so easily slips into a general Western conception of realistic depiction and aesthetic qualities – our way of depicting simply is *the* natural way. But, as the example from Australian rock art shows, to get one's eye in is always a cultural process, and as such located in historical time and space, producing not an image of what is really out there, but an image that, for being seen at all, presupposes the acquisition of certain concepts and habits of looking. Thus, there is no natural, primordial way of seeing – all that we see, all that we can see, is always already incorporated in one, ever-changing symbolic form or another.

Today, the relative dominance within cave art studies of these two approaches (which, in a sense, are but different versions of each other)[11] – that is, the longing for an origin and the mimetic curse – has produced , in spite of the many promising efforts made within the field during the structuralist era, a quite fruitless situation.[12] It is as if all contemporary attempts at making sense of what is to be found in the caves are condemned to work within this restrictive conceptual framework, where figurative depiction is seen as the natural reflection of the real. This is, of course, a typical platonic situation, excluding all kinds of constructivist approaches from the start.[13] Those who do not want to comply with this situation – and there are some – run the risk of being expulsed to the margins of the discipline, defined as freaks roaming the outside of its discursive field, as Michel Foucault might have put it.

This is where Ernst Cassirer and his philosophy of technique, technology and symbolic forms can show its worth. To approach mural cave art from the perspective of a philosophy of symbolic forms is to circumvent the urge for origin, replacing it with an immanent, productive and constructivist approach and to rethink perception in a non-mimetic way – without abandoning the demands of scientific rigour.[14] Cassirer writes:

> The symbolic process is like a single stream of life and thought which flows through consciousness, and which by this flowing movement produces the diversity and cohesion, the richness, the continuity, and constancy, of consciousness.[15]

The two key points here are, of course, the union of life and thought – that they are not to be understood as separate 'units', the one eventually being the expression of the other – and that it is by a constant flowing movement that this unity becomes productive. One of Cassirer's more perspicuous followers, Allister Neher, sums up what Cassirer tries to accomplish:

> What Cassirer attempts to do is extend the Kantian critical analysis of empirical knowledge to, as he would put it, all of the other ways human beings can have a world. In other words, he investigates the conditions of possibility for such greater cultural forms as myth, natural language, art, and, of course, contemporary science. In its most general sense, the term 'symbolic form' refers to these greater cultural forms. As organs rather than mirrors of reality, these structures are not static; unlike Kant, whose analysis of empirical knowledge assumed that Euclidean geometry, Aristotelian logic, and Newtonian mechanics were completed achievements, Cassirer builds transformation and development into his analyses of the various symbolic forms.[16]

But to see cave art only as yet another possible symbolic form is still too general an approach. It provides us with an immanent and procedural conceptual framework; it teaches us that everything, from words to things, from actions to works, that make sense to us humans, is always already meaningful within one symbolic form or another; and that the identity, as well as the sense of the objects or the actions, is thus provided by the ever-shifting, but still specific, place that it occupies within the symbolic form – or the symbolic forms in plural, since each 'symbol' can always in principle be 'grafted' into any symbolic form and hence, simultaneously or successively, incarnate different meanings.[17] Adopting this stance takes us far beyond most traditional approaches to cave art, and into a definitely post-postmodern scientific landscape – but it does not tell us much about how, or in what direction, we ought to move in our attempt to reproduce the sense of Paleolithic art. (And to reproduce is, in this context, of course not to 'collect' or mimetically 'reflect' a sense or meaning that is hiding in the depths of time – in itself an impossible task, as the history of the different attempts to do just this shows us – but to produce a sense that can be acceptable as a scientifically founded hypothesis as to what cave art once meant and could mean.) We need to be specific – focus upon Cassirer's words about the unity of life and thought and, more specifically, upon how he develops this unity in his discussion of technique, tools and technology.

Hallway and passages: Cassirer on tools and technologies

In the following sections, I would like to produce an argument for seeing mural cave art as a tool in Cassirer's sense – a move that, as we shall see, will make it possible to address the question of the sense of the art in a way that, as of yet, has not been tried in cave art studies.[18] This argument will be divided into two phases. First, I will present and, to some extent, discuss what I take to be the most relevant aspects of Cassirer's take on technology in relation to the problems evoked in and by the study of mural cave art. In a second move, I will present, in a very condensed way, a concrete experiment conducted by the French cave art specialist Michel Lorblanchet in the Pech Merle cave in the department of Lot, France. Finally, I will contend that what goes for cave art, goes for all tools, conceptual as well as technological, claiming that what we may learn in the caves echoes in contemporary science.[19]

In the second volume of his *Philosophy of Symbolic Forms*, devoted to mythical thought, Cassirer writes a few pages on technique, tools and technology. In a seminal passage, predating his 'Form and Technology' by five years, he writes:

> Thus we see that even if we regard the implement purely in its technical aspect as the fundamental means of building material culture, this achievement, if it is to be truly understood and evaluated in its profoundest meaning, may not be considered in isolation. To its mechanical function there corresponds here again a purely spiritual

function which not only develops from the former, but conditions it from the very first and is indissolubly correlated with it. Never does the implement serve simply for the mastery of an outside world which can be regarded as finished, simply given 'matter'; rather, it is through the use of the implement that the image, the spiritual, ideal form of this outside world, is created for man.[20]

What I find remarkable in this passage is, again, the way Cassirer connects life and thought as embodied in man's tools, seeing these tools as productive of man's world in a simultaneously both material and mental sense. And he chooses, approvingly quoting Ernst Kapp and his *Grundlinien einer Philosophie der Technik* from 1880, the term 'organ-projection' as an adequate description of this process, indicating the possibility of a two-way movement – from organ, primarily hand, to 'hammer, hatchet, ax, knife, chisel, drill, saw, and thongs' but also – and this is the crucial point here – from tools and technology to the human being, that is 'by reversal of the process as a means of explaining and understanding the human organism'.[21] In this passage, Cassirer is only talking about tools in an ordinary sense (hammer, knife chisel), but for me the term 'organ-projection' immediately connects to the projected, negative and positive, hands we find in the Paleolithic caves; and subsequently to all other traces found in the caves. Seeing Paleolithic mural cave art as a multiple tool in this sense – that is, as a symbolic form creating a world for man, seeing it as an organ 'less of mastery than of signification' (as Cassirer puts it) – seems like a very plausible and possibly fruitful approach not explored within the confines of the discipline of cave art study.[22] It has the great advantage of providing a conceptual framework that allows us to ask not *why* men painted, engraved or sculpted in the caves, but what kind of sense-making the sculptures, the engravings and the paintings allow of today – that is, what kind of scientifically acceptable sense we can possibly produce about the practices and, perhaps, the general attitudes that once engendered the traces.[23]

Turning to Cassirer's more developed arguments in 'Form and Technology', there is further support for this view, most significantly perhaps in the way that Cassirer develops, again, the notion of co-determination between life and thought.[24] He writes, for instance, about the intermingling and confrontation ('Auseinandersetzung') of technology and all other 'areas and entities'[25] of human life:

Thus, every introduction of a new element not only widens the *scope* of the mental horizon in which this confrontation takes place, but it alters the very *mode of seeing*. This formative process does not only expand outwardly – it itself undergoes an intensification and heightening, so that a simultaneous qualitative transformation occurs, a specific metamorphosis.[26]

Through this insistence on the principle of becoming, of creating and of transformation, adopting Cassirer's stance in relation to cave art precludes the traditional ways of approaching the traces in the caves as static reports from and depictions of a world long lost. Instead, Cassirer opens the possibility of seeing the traces in the caves as tools that, if we just could learn to use them, would let us recreate if not a whole mental universe, then at least some of the gestures and fragments of moves that perhaps constituted the practices that these traces once were living parts of – the idea being, once again, to move from the tools towards their possible significations, as eventually revealed in both the mental and bodily practices they are the traces of. Cassirer's own focus seems to coincide with my own here. He writes:

If, instead of beginning from the existence of technological objects, we were to begin from the technological efficacy and shift our gaze from the mere product to the mode and type of production and to the lawfulness revealed in it, then technology would lose the narrow, limited and fragmentary character that otherwise seems to adhere to it. Technology adapts itself – not directly in its end *result* but with a view to its task and *problematic* – into a comprehensive circle of enquiry within which its specific import and particular mental tendency can be determined.[27]

Of course, in the case of cave art it is no longer possible to focus directly upon the practices of producing paintings in the caves – no matter what the proponents, still found in ethnology, anthropology and archaeology, of the erroneous (in my eyes, at least) 'argument through analogy' may claim.[28] But Cassirer opens another possibility when he talks about the mode, the type of production and the lawfulness that is revealed in the work. As we will see below in the discussion of Lorblanchet and his experiment, this possibility has been partly explored – and with surprisingly interesting results. It seems as if Cassirer was right in not upholding strict, absolute borders between technology and language or between tools and artworks – through their always being part of some symbolic form, they all partake in the human sense-making; and seen from this perspective there is no strict difference between language, art and tools. In fact, Cassirer claims that:

> It is not the human being who, as mere natural being, as a physical-organic being, becomes the creator of art; rather it is art that proves to be the creator of humanity, that first constitutes and makes possible the specific 'mode' of being human.[29]

So, in this specific sense, art may definitely be seen as a tool, allowing us, at least in principle, to recreate the sense and lawfulness of its production.[30] And if this is the case, it should not be impossible to move from the cave art to the 'organ' that projected it – 'organ' here being understood as the symbolic form in which the traces in the caves once functioned – as when Cassirer, with a reference to Kapp, claims that all effective technological activity 'likewise always exhibits a self-revelation, and, through this, a means of self-knowledge'.[31] In the case of cave art, it is up to us, researchers of today, to find the means and the knowledge necessary for using this revelation in making scientifically tenable sense of the traces in the caves.[32]

Galleries: on the remaking of the horses in Pech Merle

The already-mentioned French researcher Michel Lorblanchet is among those who have tried to do just this, coming very close to a Cassirerian perspective. Focusing on the practical, material and physical aspects of the art of the upper Paleolithic era – such as, for instance, what colours were used? what tools? which techniques were instrumental in producing what effect? how does the environment in the cave interact with the paintings? and so on – he has clearly proven the value of embodied experimentation in cave art studies.

His well-documented life-size remake of the panel of horses in Pech Merle is significant in many ways – not least for the role it has played in refuting the central tenets of the shamanist theory, such as the presumed universal identity of trance-experience allegedly caused by the way the human nervous system works; the presumed importance of the very act of painting; and the toxicity of the dyes used, supposedly inducing hallucinations and trance states in the painters.[33] Important as this may be, what is central in this experimentation from our point of view is what we may learn today about the why and wherefore of the horses by coming to grips with how they were made in the first place. Is it possible to gain any insight as to their signification from a remaking such as the one made by Lorblanchet? Such a remaking will never, of course – and I want to stress this to avoid any possible confusion – provide an access to intentions long lost, or to the original sense of the cave art,[34] but it may provide an as of yet unused road to follow in the search for a scientifically acceptable way of construing the sense of the traces in the caves. In a moment, we will look at Lorblanchet's own conclusions. But first, a very brief description of what he did.[35]

Lorblanchet started by making a copy (a *relevé*) of the original panel, using both tracing and infra-red photography. Then he looked up a suitable surface on which to execute the copy in a non-painted cave, near the original one. Then he reproduced the pigments: the black was mainly made out of charcoal from oak and juniper, chewed and mixed with saliva, using water as dilution;[36] the red was ochre found near the cave. Based on the information obtained, he was able to establish that the panel was constructed in four major phases, starting with the horses and ending with the pike and the other motives in red. (This order was later revised by Lorblanchet himself, making the pike the first phase out of five.[37]) Analysing the phases one by one, he was able to identify and reproduce the different techniques used, such as spitting the colour

Figure 10.1   The Panel of the Horses, Pech Merle, France. © Michel Lorblanchet

directly onto the surface (*crachis*), using a hollow bone for blowing the colour onto the surface (*flocage*) and also how to use his hands as templates. To obtain an acceptable result for the black points, he used a piece of leather with a round hole in the middle, thus protecting the surface from surplus pigments produced by the spitting procedure.[38]

The results obtained in the process were, as I have already said, quite astonishing: Lorblanchet could confirm that the techniques he used (*crachis* and *flocage*), techniques that he initially learned from aboriginal inhabitants in Australia, must have been the ones used when making the original horses; he could make an estimation of the total time for making the panel: 32 hours – that is, much longer than it would have taken using other techniques, such as brushing, or using stamping techniques. He could also conclude that the horses must have been sketched in outline before being painted – during the procedure of *crachis* the mouth has to be very close to the surface, making it impossible to maintain an overall view while working – adding to the argument that painting in the caves was a meticulously planned and structured activity and hardly something that could, at least when it comes to the more elaborated paintings or engravings, be executed in trance, or while having hallucinations, or just as a momentary impulse.[39] Thus, guiding lines were necessary so as not to lose touch with the general composition and start making mistakes. Moreover, he could prove that there had to be at least two people that were continuously involved in the execution of the panel – one painting, the other providing illumination by holding grease-lamps close enough to the surface. Lorblanchet was also able to show that the entire panel was most likely executed at the same time and by the same artist – and that it has been renewed at different intervals, long after its completion, a circumstance that (to say the least) indicates a lasting interest in the panel. Commenting upon the method of experimentation, Lorblanchet writes:

> The principal aim of making this kind of mural experiments is to find the intention behind the gesture, the 'driving force' behind the act or, rather, to better understand the mind, and sometimes the symbolic context, within which the works were accomplished. The choice of cavern, of stony support and of pigments; the time of execution; the composition of the panel; the working-conditions for the artist (accompanied or solitary); the will to exhibit or to hide the work; the usage of the extant shapes of the cave wall; the verification of aesthetic mastery etcetera make up the field of experimentation. This information is indispensable and partakes in the interpretation of paleolithic art.[40]

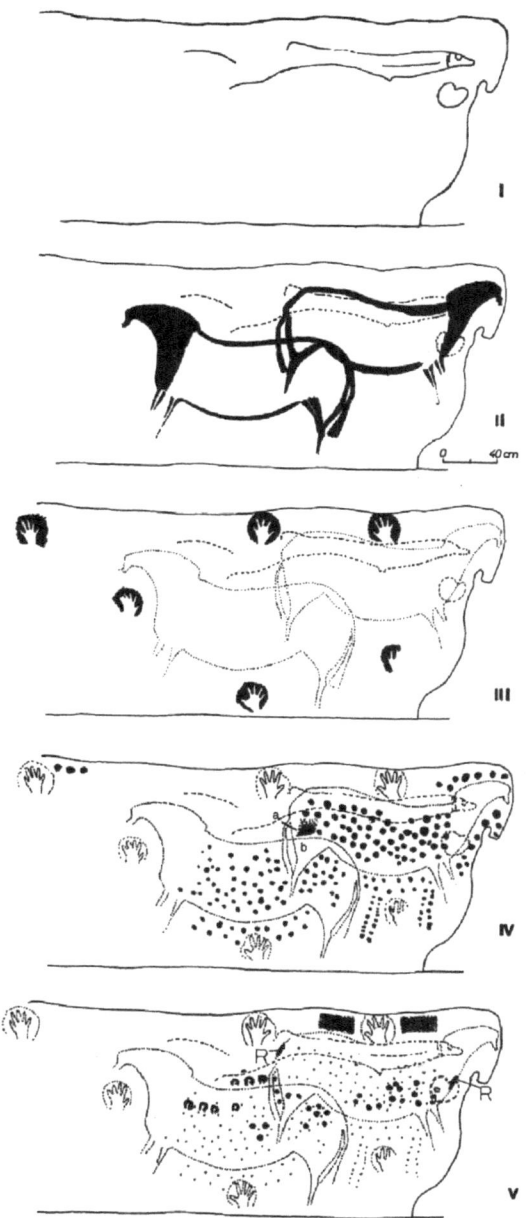

Figure 10.2 The phases of construction of the Panel of the Horses. © Michel Lorblanchet

But even though Lorblanchet is explicitly looking for the intention behind the gesture (an approach that I find rather misdirected), he does not make the mistake of advancing a general hypothesis about the sense of the cave art – probably wisely warned by his familiarity with all previous and vain attempts. Experimentation can only provide probabilities, not proofs, he says. It can 'underpin a hypothesis by showing not the reality of the anticipated facts, but their *makeability* (faisabilité)'. But at the same time he claims, more in accordance with Cassirer than he is probably aware of, that the technique in itself is 'loaded with meaning: there is not, of course not, any gulf separating technological research from research after meaning'.[41]

Exit: on the way out of an impasse?

So, in what ways may Cassirer's notions of tools, technology and symbolic forms help us in our effort to make sense of the Paleolithic art? In trying to answer this question, we have to be both wary, curious and creative. Wary, so as to avoid the pitfalls common to all attempts at explications of the sense Paleolithic art – after all, we are dealing with art that was produced for a period of more than 20,000 years, during periods of severe glaciation as well as in more benign conditions. What are our claims for talking about a single kind of art during this immense span of time? What would it imply as to the stability of society and culture, for instance? It is a fact that we tend to create the art of the upper Paleolithic as a homogeneous unity through inscribing it in, as it were, the symbolic form art, with all its aesthetic and formal concerns – and such an inscription is perhaps an explainable move, given the cultural context of the discovery of cave art, but it is hardly a necessary one. In any case, it does not offer us any support for claiming that man has an innate desire or capacity for beauty, or urge for making art, as so many have done and still do when trying to fathom Lascaux, Chauvet or Niaux.[42]

   Minding at least these pitfalls, we still must be curious and creative enough to try, in spite of previous failures, to circumvent the problems and find a way out of the dead-end announced by the recurring mantra 'we will never know'. As a first response, I would say that whether we will ever come to know anything about the meaning of the cave art depends entirely upon what we mean by 'knowing' in this context. We will, as already said, never be able to enter into the minds of the painters and pinpoint their intentions. But, then again, this does not, in principle, make cave art any different from trying to understand other kinds of symbolic expression, from any time whatsoever. The difference, and this is of course an important difference, lies in the absence of a scientifically acceptable and sufficiently precise context (that is, a context that, by and large, is accepted as relevant by those who are working in the epistemic fields of archaeology, anthropology, ethnology, etc.) that we may use as a framework when trying to interpret works of Paleolithic art. Such a context is still, despite the vast knowledge that we actually possess about our prehistory, desperately lacking. So this context has to be created. And I would suggest that the thoughts of Cassirer may be instrumental in this work. His notions and understanding of symbolic forms, tools and technology will not give access to the sense of the paintings in the caves, but they may well direct our sense-making efforts in a new way, and perhaps also provide the beginnings of a context for the paintings.

More concretely, this would at least mean refusing the mimetic/realistic curse or paradigm in cave art studies by trying to move 'from the mere product to the mode and type of production and to the lawfulness revealed in it'[43] – that is, to try to reconstruct the symbolic form that is revealed in the paintings in the caves. This approach is, of course, not entirely new – the first attempts at making sense of cave art consisted in trying to make it comply with totemism, sympathetic magic and mythological thinking, as well as, from the late 1950s onward, with an enlarged structuralist conception of language.

But all these approaches started in the wrong place, so to speak: they all assumed that cave art would fit into some or other of the extant forms of human culture. Cassirer invites us to do exactly the opposite: start with the work or works and see what they reveal in and through themselves; do not presuppose anything specific about them from the start – you will anyhow inevitably bring along enough culturally dependent presuppositions, functioning as the inescapable conditions for your own sense-making. The remaking of the Panel in Pech Merle may serve as one example of how such a work may be started; there are no doubt other ways of proceeding as well. For instance, focusing even more than Lorblanchet did on the very gestures involved in painting, using the concept of 'organ-projection' as a tool for moving from gesture to subjective, bodily experience, may teach us surprising things about the limits of makeability and understandability. If a gesture involved in the making of a painting seems physically awkward or strange, this may, for example, indicate that there is something more to the picture than meets the eye.

Cassirer's philosophy also urges us to see the paintings in the caves not as ready-made reality, but as traces of a world in the making. And such a shift in perspective is no doubt decisive for our sense-making – we are no longer forced to look for a forever-forgotten meaning hidden deep within or behind the symbols; instead we are offered the possibility of putting the traces we meet into play, and seeing what they may start to generate. This approach is not entirely new either: researchers such as, most famously, Annette Laming-Emperaire and André Leroi-Gourhan have attempted similar approaches, trying to make the caves reveal their syntax and semantics through activating a grid of binary oppositions into which they tried to fit the paintings, symbols and traces.[44] But they did not really pay heed to the embodied aspects of sense and meaning, highlighted by Cassirer, and thus mistook the impossibility of integrating all traces into one grid for a failure. The notion of embodied meaning forces us to adopt a more specific, physically as well as culturally situated, way of dealing with the caves and the paintings – what goes for the horses of Pech Merle is not necessarily valid for the panel of lions in Grotte Chauvet. Each work has to be seen in its specificity, as a tool designed to do something specific in just this location, otherwise we no doubt will miss important details and hints. But the work would not be a tool if it was not also a part of a more general symbolic form, involving practices and beliefs, techniques and intentions. Cassirer's take on sense and symbols, as well as the work of experimental researchers, makes it possible to look at this tension between specificity and universality in, at least in cave art studies, a more or less unexplored way.

Outside again

I hope that I now have managed to show some of the topicality of both Cassirer and of cave art: Cassirer's philosophy of technology and symbolic forms is consonant with many important modes of contemporary thought – from the current interest in immanence and stratification to the latest advancements in cognitive science, passing through political science and political philosophy, the notions of embodiment and becoming are central today, and in need of clarification and development. Both can be found in Cassirer's works, once you start to read him as a contemporary – or, perhaps, as a doxological thinker *avant la lettre* – and not only in order to establish his position in the canons of twentieth-century philosophy. And to end this cavernous chapter, the case of cave art studies may, I hope, serve as an example of how we may proceed in creating, in the guise of recreation, a symbolic form consonant with the specific epistemic needs of our current scientific and political situation, where bodies have come to matter more and more, in every kind of sense.

Notes

1. Since 2002 I have been developing an 'other', rhetorical take on epistemology. I call this stance doxological in order to emphasize that all knowledge is doxic knowledge, thus turning the seminal Platonic distinction between doxa (beliefs, opinions) and episteme (objective, eternal knowledge) upside down. Protagoras dictum about man as the measure of all things is, perhaps, the most poignant expression of a doxological position. Departing from the pivotal question 'What would a Protagorean position imply for epistemology today?', I propose a critique of the purely discursive notion of knowledge, still central in Anglo-Saxon epistemology, emphasizing the fact that our knowledge is always embodied – in ourselves as biological beings; formulated and/or preserved in some language, institution or ritual; practised and upheld by one or many individuals, always in one historical moment or other and within the admittedly diffuse framework of an ever-changing but still specific social situation. Doxology is not a relativism abandoning all claims to objectivity or science – far from it – but an attempt, in the wake of the serious and fundamental criticisms of the late twentieth century, to readdress and reconsider how we can conceive of knowledge, science and objectivity today. Please also note that I use the term 'doxology' in a completely non-theologian sense, as a designation for an epistemological stance, exploring not episteme, but doxa. See Mats Rosengren, *Doxologi: En essä om kunskap* (Åstorp: Rhetor, 2002), in French translation as, *Doxologie: Essai sur la connaissance* (Paris: Hermann, 2011).
2. For an historic overview of the different attempts at interpreting cave art, see, for example, David Lewis-Williams, *The Mind in the Cave* (London: Thames and Hudson, 2002), and Anati Emanuele, *Aux Origines de L'art* (Paris: Librairie Arthème Fayard, 2003).
3. That is – no tradition that we know anything of today. In the 'youngest' of the caves where Paleolithic art-objects have been found known today – La Grotte de Mas d'Azil – the most recent findings are about 10.000 years old (counting backwards from 1950, the year of reference in when it comes to dating in archaeology), but to talk about either Mas d'Azil, Lascaux or any other painted cave as the origin of art, or even of Western art, is simply strange. As far as we know today there is no continuity or even connection between the Paleolithic mural art and other purported origins of Western art, be it ancient Egypt or ancient Greece. I think that we, in the case of art as elsewhere, have to content ourselves with multiple origins, and with an ongoing creation. (Cornelius Castoriadis has devoted many books to the problems of human creation, and what this capacity of ours entails – see, especially, Castoriadis, *La Creation Humaine* (Paris: Seuil, 2002).)
4. John Clegg, 'Pictures and Pictures of …', in Paul Bahn and Andrée Rosenfeld (eds), *Rock Art and Prehistory Papers Presented to Symposium G of the Aura Congress, Darwin 1998* (Oxford: Oxbow Monograph 10, 1991), 109–10. Even though I find the comments on perception just quoted quite convincing, I do not share Clegg's general approach (that the problem is only one of vocabulary) nor his conclusion (111) (that 'there is a need for a convention to indicate whether a term is in use as a name or label') presented in his article.
5. Jean Louis Schefer, *Questions d'art paléothique* (Paris: POL, 1999). All translations are my own, unless otherwise indicated.

6. *Ibid.*, 24. I have discussed the problems involved in using both the term and the concept 'art' in connection to the findings in the prehistoric caves in 'The Cave of Doxa: Reflections on Artistic Research and on Cave Art', in *Art Monitor*, 4 (Faculty of Fine, Applied and Performing Arts, Gothenburg University, 2008).

7. Jean Clottes and David Lewis-Williams, *Les chamanes de la préhistorie: Texte integral, polemique et réponses* (Paris: La maison des roches, 2001), 45.

8. *Ibid.*, 45–6.

9. See my 'On Creation, Cave Art and Perception: A Doxological Approach', *Thesis Eleven*, 90:1 (2007), and also my *Cave Openings: On Cave Art, Perception and Knowledge* (forthcoming).

10. One may even say that the discipline of cave art was constituted around this claim. This is how Eduard Harlé described the paintings in the Altamira cave, arguing, like almost everyone else in the scientific community at the time (early 1880s), that they were forgeries: 'All trace of smoke coming from the fireplaces having disappeared, why then should the paintings remain? Since the light of day does not reach the emplacement of the pictures, and since their making must have taken a certain time, they could only have been made in an epoch when sophisticated artificial lighting could be provided. Moreover, the paint comes off easily, it covers several stalactites, and when examined proves to be the product of learned processes ... But above all, unlike the very realistic hind, the aurochs that the maker has taken the trouble to paint do only partially represent the traits of this animal; therefore he has never seen them.' Quoted after Marc Groenen, 'Introduction' and 'Presentation Générale', in André Leroi-Gourhan, *L'art parietal: Langage de la préhistorie* (Grenoble: Editions Jérome Millon, 1992), 71. It is, of course, the very last sentence in this quote that I would like to highlight in this connection, since I see it as a direct expression of the mimetic curse that I am talking about here. Harlé's way of arguing against the authenticity of the paintings was accepted as valid up until 1901– but the idea that, in order to depict in a realistic way, the painter must have seen the object depicted 'in real life' still lingers in cave art studies.

11. Both are, in fact, versions of what Cornelius Castoriadis would call 'the myth of being as determined', and represent examples of the heteronomic way of thinking that he so resents. For explorations of Castoriadis' philosophy from this perspective, see my *För en dödlig, som ni vet, är största faran säkerhet: Doxologiska essäer* (Åstorp: Rhetor, 2006), and my 'Radical Imagination and Symbolic Pregnance: A Castoriadis Cassirer Connection', in *Embodiment in Cognition and Culture*, ed. Krois *et al.* (Amsterdam: John Benjamin, 2007).

12. All that I have said this far is but a very thin and reductive description of a very complex scientific field and of a multilayered problematics – there is, of course, a lot more to be said about the discipline of cave art, the history of the development of different interpretations and about the different aesthetic, practical, theoretical and scientific approaches to the caves and what is found in them. For an extensive overview, see Lewis-Williams, *The Mind in the Cave*. For a critical presentation, see my 'On Creation' and my 'The Cave of Doxa'.

13. I have spelled out this platonic analogy in some detail in the first chapter of *Cave Openings: On Cave Art, Perception and Knowledge.*

14. In my works on doxology, I have used the ideas of Ludwick Fleck to present how such a constructivist and yet rigorous approach may be formulated, using Fleck's concepts of 'thought collective', 'thought style' and 'active and passive connections'. As I see it, Cassirer's notion of symbolic form incorporates and in some ways furthers Fleck's approach. But this is not the place for developing this comparison. See my *De symboliska formernas praktiker: Ernst Cassirers samtidiga tänkande* (Göteborg: Art Monitor Essä, 2010). See also note 23, below.

15. Ernst Cassirer, *The Philosophy of Symbolic Forms*, vol. 3: *The Phenomenology of Knowledge*, trans. Ralph Manheim (New Haven: Yale University Press, 1957), 202.

16. Allister Neher, 'How Perspective could be a Symbolic Form', *The Journal of Aesthetics and Art Criticism*, 63:4 (2005), 360.

17. I have argued for the compatibility of Jacques Derrida's notion of 'iterability' and 'quotability' and Cassirer's notion of 'symbolic pregnance', in my 'Radical Imagination'.

18. At least not in explicitly Casserian terms, I should perhaps add. As we will see below in the discussion of Lorblanchet's experiment, at least one attempt in this general direction has been made.

19. A terminological note: When I use the term 'science' in this general way, I include the humanities as well as the social and natural sciences. This is no doubt a bit odd for a native English speaker, but I want to insist on this usage, as it makes a case for the epistemological equivalence of the different branches of science.

20. Cassirer, *Philosophy of Symbolic Forms*, vol. 3, 215.

21. *Ibid.*, 215–16.

22. *Ibid.*, 217.

23. The words 'scientifically acceptable' are crucial here. They indicate the doxological preconditions, formulated in terms of thought collective, thought style, active and passive connections and epistemic domain, for our production of knowledge. See Rosengren, *Doxologi*.

24. As the text in question is included and discussed elsewhere in this volume, resuming its main arguments seems a little presumptuous. Thus, I will only highlight some passages that I find specially rewarding, considering my intentions in this essay.

25. See Ernst Cassirer's 'Form and Technology', in this volume, 17.

26. *Ibid.*, 17–18.

27. *Ibid.*, 20.

28. The 'argument through analogy' is based on the idea that we can understand the way of living and thinking during Paleolithic times by comparing with so called 'primitive' cultures living in more or less the same way as we think our forbears did. For a critical discussion, see chapter 6 of my *Cave Openings*.

29. Cassirer, 'Form and Technology', 36.

30. I have already pointed out that there are major problems and conceptual difficulties involved in talking about what we find in the caves as art. For an extensive discussion, see Rosengren, 'The Cave of Doxa'.

31. Cassirer, 'Form and Technology', 38.

32. I do not think that the difference between contemporary artistic and technical creation and beauty that Cassirer discusses at length towards the end of

his text, a discussion that in itself would merit a careful and subtle analysis, invalidates my argument concerning cave art as a tool – cave art being, after all, one of mankind's very first symbolic formations, preceding the distinction of techne and physis by some 35,000 years.

33. I have discussed the shamanist theory at some length in 'The Cave of Doxa'. For Lorblanchet's own development, see especially Lorblanchet's own article 'Rencontres avec le chamanisme', in Michel Lorblanchet, Jean-Loïc Le Quellec, Paul Bahn, and Henri-Paul Francfort (eds), *Chamanisme et Arts Préhistoriques: Vision Critique* (Paris: Errance, 2006).

34. As I hope will become clear, the value of experimentation does not lie in a purported recreation of original intentions, nor in a creation of a more true (whatever that would mean in this context) historical meaning: no, its value and potential lies in the way experimentation allows us to affirm the connectedness of mind and body in the process of sense-making today – as, for example, when establishing scientifically tenable hypotheses about the different possible senses of cave art.

35. The full story is presented in Michel Lorblanchet, *Les grottes ornées de la Préhistoire: Nouveaux regards* (Paris: Errance, 1995), 209–23 and partly updated in Lorblanchet *et al.*, *Chamanisme et Arts Préhistoriques*.

36. Lorblanchet has a long argument concerning the eventual effects of toxicity of the managnese oxide, a constituent of the original dyes, in his 1995 book. This argument is developed into an outright critique of the shamanist thesis in Lorblanchet, 'Recontres avec le chamanisme' – but I leave this debate aside, since it is of no direct interest for our main concerns here. The locus classicus for a presentation and arguments for the shamanist theory is Clottes and Lewis-Williams, *Les chamanes de la préhistoire*.

37. See Lorblanchet *et al.*, *Chamanisme et Arts Préhistoriques*, 107.

38. Lorblanchet, *Les grottes ornées*, 210–16.

39. It seems quite probable that not all art in the caves served the same purpose – for example, big paintings in accessible areas (such as the big hall in Lascaux) were probably intended to be seen by more people than just the painters, and were perhaps also used in rituals of one kind or another; other paintings and engravings are obviously hidden in the most remote and hard-to-get-to parts of the caves, indicating another usage and significance.

40. Lorblanchet, *Les grottes ornées*, 222.

41. *Ibid.*, 223.

42. A recent example is Gregory Curtis, *The Cave Painters: Probing the Mysteries of the World's First Artists* (New York: Alfred A. Knopf, 2006).

43. Cassirer, 'Form and Technology', 20.

44. See, for example, Annette Laming-Emperaire, *La Signification de l'Art Rupestre Paléolotique* (Paris: Editions A. et J. Picard, 1962) and Leroi-Gourhan, *L'art pariétal*.

11
Failures of Convergence

Dennis M. Weiss

I

Ernst Cassirer opens his 1944 *An Essay of Man* arguing that while self-knowledge is the highest aim of philosophical enquiry, today man's knowledge of himself is in crisis. Cassirer points out that no former age was ever in such a favourable position with regard to the sources of our knowledge of human nature. As he notes: 'Psychology, ethnology, anthropology, and history have amassed an astoundingly rich and constantly increasing body of facts. Our technical instruments for observation and experimentation have been immensely improved, and our analyses have become sharper and more penetrating.'[1] And yet, Cassirer argues, we have no method for the mastery and organization of this material. We have a mass of disconnected and disintegrated data which seem to lack all conceptual unity. The anarchy of thought, Cassirer notes, leaves us without a frame of reference or general orientation and our wealth of knowledge threatens to become little more than a mass of disconnected and disintegrated data. This, Cassirer notes, is a danger, a theoretical as well as a practical problem. As he writes: 'That this antagonism of ideas is not merely a grave theoretical problem but an imminent threat to the whole extent of our ethical and cultural life admits of no doubt.'[2] And on this point he cites Max Scheler who notes that 'in no other period of human knowledge has man ever become more problematic to himself than in our own days'.[3] Cassirer's later work, including *An Essay on Man*, *The Myth of the State* and many of the essays collected in *Symbol, Myth, and Culture*, is particularly imbued with an awareness of 'menacing danger',[4] as he refers to it in the essay 'The Concept of Philosophy as a Philosophical Problem', and the slow disintegration and the sudden collapse of social and political life in the last decades,[5] as he puts it in the essay 'Judaism and the Modern Political Myths', written in 1944, the same year as the publication of *An Essay on Man*. Cassirer worries that modern philosophical thought has become increasingly pessimistic and fatalistic and that philosophy has abrogated its ethical responsibility to speak to these theoretical as well as practical crises.

Cassirer's reference in *An Essay on Man* to Max Scheler is particularly noteworthy in this context as Scheler's 1928 work *Man's Place in Nature*, from which Cassirer quotes, is widely regarded as the work initiating the German tradition of philosophical anthropology. For both Scheler and Cassirer, the way out of this crisis is a renewal of anthropological thought and, indeed, within the tradition of philosophical anthropology, Cassirer's notion of a crisis in self-knowledge and the need to situate that crisis within anthropological thought was quite common. Similar sentiments can be found in the philosophical anthropologies of Arnold Gehlen, Helmuth Plessner and Michael Landmann. In *Man: His Nature and Place in the World*, Gehlen observed that 'Man is a being whose very existence poses problems for which no ready solutions are provided.'[6] And Michael Landmann had this to say in *Philosophical Anthropology*; 'Man has become problematic as never before; he no longer knows what he is, and he knows that he does not know it.'[7] In his 1954 book *The Social Self*, Professor of Philosophy Paul Pfuetze eloquently gave voice to this sentiment, which he found widespread in twentieth-century culture. 'There is', he writes:

a crisis and revolution in modern culture and in man's knowledge of himself which has occasioned a revival of interest in anthropology both in philosophical and in theological circles. Modern man has become a problem to himself, and all over the world men are inquiring with fresh zeal into the nature of man. What is man? What is the meaning of human existence? In the confusion of voices, a deep disquietude has fallen upon the human race. On all sides one finds moral disaster, political confusion, spiritual discontent, mental breakdown, and a growing suspicion, now amounting to a certainty, that during the last few centuries man has so far misinterpreted his own nature as to make tragic and catastrophic use of his powers and technics.[8]

There is, Pfuetze notes, the growing suspicion that until now we have got things wrong, misinterpreted our nature as human beings. And this misinterpretation is, at least in part, responsible for the tragic and catastrophic misuse of our powers and technology. In 'Form and Technology', Cassirer as well references this awareness of our misuse of our powers and technology. He raises concerns over technology's breadth and growing power,[9] its subjugation of modern culture, and its role in severing the human being from organic life. He quotes philosopher Ludwig Klages' observation that the human being is possessed by technology, a vampiric and soul-destroying power[10] and worries over the grave inner damages of a technological culture which throw the human being into a 'never-ending vertigo'.[11] Evident in 'Form and Technology' is the debate, already extensive by 1930, over the impact of the primacy of modern technology – whether it should be blessed or cursed and whether it is a source or symptom of the crisis facing culture.

In light of this crisis of self-knowledge, Cassirer and these other figures turned to philosophical anthropology as a path out of the crisis. Each raises the anthropological question: what am I that I am a human being? Pfuetze speaks of the fresh zeal with which scholars were turning to the question of the nature of the human being, a zeal that brings with it a renewed hope that a new form of anthropological thought may rescue us from moral disaster, political confusion, spiritual discontent and mental breakdown. Beneath the sense of crisis and catastrophe, then, lies the hope that if we can properly answer this question, if we can approach the anthropological question without misinterpretation, then perhaps we might avoid these mistakes, or at least go some way towards correcting them. For Cassirer too, as we shall see, the 'clue of Ariadne' which will lead us out of this labyrinth lies in a fresh approach to the anthropological question and a recognition of man's symbolic nature.

Fast forward some 50 years, though, and it is precisely the anthropological question, Cassirer's question regarding self-knowledge, that is rendered problematic by the advance of our powers and technology, for those very powers and technologies are now being turned back on the human being. Developments in genetic engineering, biotechnology, neuro-pharmacology, robotics and prosthetics raise the spectre that the human being itself may be refashioned and re-engineered. Indeed, in setting out to describe our current situation, it is clear that there is a widespread presumption that humanity may be at a turning point. Issuing from a variety of perspectives and motivated by a cross-section of theoretical concerns, comes the claim that, especially owing to technological developments, human beings are on the cusp of profound change. For instance, Susan Squier, Brill Professor of Women's Studies and English at Pennsylvania State University, notes in *Liminal*

Lives that biomedicine and biotechnology are reshaping our ways of conceiving, being born, growing, aging and dying, changing the expected shape and span of human life. 'The foundational categories of human life have become subject to sweeping renegotiation under the impact of contemporary biomedicine and biotechnology.'[12] Duke University Professor of Literature Katherine Hayles agrees, suggesting that 'technology has progressed to the point where it has the capability of fundamentally transforming the conditions of human life'.[13] In his recent essay 'Icarus 2.0', historian Michael Bess argues that we are in the early stages of an 'epochal shift' that will prove as momentous as such great transformations as the transition from hunting and gathering to settled agriculture and the substitution of steam power for human and animal energy. We are, he suggests, at a turning point that will shake ethical and social foundations, as we apply the technologies of human enhancement (which he identifies as pharmaceuticals, prosthetics/informatics and genetics) to the reinvention of our own physical and mental capabilities. 'Though advances in each of these three domains are generally distinct from those in the other two, their collective impact on human bodies and minds has already begun to manifest itself, raising profound questions about what it means to be human.'[14] And like Cassirer, some 50 years earlier, Bess's essay often refers to these developments as destabilizing, dramatic and disorienting, emphasizing the sense of crisis that attends the birth of the post-human. Today, though, that crisis extends to the very question Cassirer posed in his *Essay on Man*: 'What is man?' What is man when his nature can be re-engineered through the technologies of human enhancement?

While Bess focuses primarily on biotechnology, others look more broadly at a host of technologies that are converging to radically alter human nature. In *Radical Evolution: The Promise and Peril of Enhancing Our Minds, Our Bodies – and What It Means to Be Human*, Joel Garreau, an American journalist and author, focuses on 'the future of human nature' and explores the 'biggest change in tens of thousands of years in what it means to be human'.[15] Garreau's discussion focuses on robotics, information science, nanotechnology and genetics, which he refers to as the GRIN technologies, and ponders the questions 'will human nature itself change? Will we soon pass some point where we are so altered by our imaginations and inventions as to be unrecognizable to Shakespeare or the writers of the ancient Greek plays?'[16] It is this fear that is the focus of Bill Joy's essay 'Why the Future Doesn't Need Us.' Joy, co-founder and chief scientist of Sun Microsystems, has been having second thoughts about the computer revolution and in his essay explores how it is that 'our most powerful 21st-century technologies – robotics, genetic engineering, and nanotech – are threatening to make humans an endangered species'.[17]

Joy was responding in part to the growing interest in the Singularity, the notion of rapidly accelerating technological change first popularized by Vernor Vinge and most recently the focus of inventor and futurist Raymond Kurzweil's book *The Singularity is Near: When Humans Transcend Biology*. In his earlier book, *The Age of Spiritual Machines*, Kurzweil argued that 'the primary political and philosophical issue of the next century will be the definition of who we are'.[18] In *The Singularity is Near*, Kurzweil points to a different kind of turning point and argues that the pace of accelerating change in technology will lead in the not-too-distant future to what Garreau refers to as an imminent and cataclysmic upheaval in human affairs[19] and what Kurzweil suggests will be a rupturing of the fabric of human history.[20] As Kurzweil notes:

> What, then, is the Singularity? It's a future period during which the pace of technological change will be so rapid, its impact so deep, that human life will be irreversibly transformed. Although neither utopian nor dystopian, this epoch will transform the concepts that we rely on to give meaning to our lives, from our business models to the cycle of human life, including death itself.[21]

Kurzweil predicts that the Singularity will occur in 2045,[22] just a short 100 years following the publication of Cassirer's 1944 *An Essay on Man* and his reflections on our crisis of self-knowledge. Far from Cassirer's sense of crisis, however, Kurzweil hypothesizes that these technological developments will bring a period of limitless opportunity and advancement. While he recognizes that genetics, nanotechnology, robotics and AI bring with them deeply intertwined benefits and dangers, he argues that in the end, 'it is only technology ... that will offer the leverage needed to overcome problems that human civilization has struggled with for many generations'.[23] Do developments in the technologies of genetics, nanotechnology, robotics and AI point to a way out of the theoretical and practical crises noted by Cassirer, Scheler, Pfuetze and others? Or will they only serve to deepen our crisis of self-knowledge as we confront anew our powers and technology, now focused more specifically on the human being? Recent proposals emphasizing the convergence of technologies on the improvement of human performance promise a way of out of this dilemma and it is to those proposals that I turn in the next section.

II

In the USA, 'converging technologies' refers to the synergistic convergence of nanotechnology, biotechnology, information technology and the cognitive sciences, commonly jointly referred to as NBIC technologies. For several years now, the motivating force behind discussions of converging technology in the USA has come from Mihail Roco and William Sims Bainbridge. Roco is chair of the US National Science and Technology Council subcommittee on Nanoscale Science, Engineering and Technology (NSET), and Senior Advisor for Nanotechnology at the National Science Foundation. Bainbridge, a professor of sociology, currently serves as Co-Director of Human-Centered Computing at the National Science Foundation. They have organized a series of workshops and conferences on converging technologies and published proceedings from the first four workshops.[24]

Following a quick perusal of the published proceedings from the converging technologies workshops, one might be forgiven for thinking it a mistake to turn to a discussion of NBIC convergence in the midst of a discussion of crisis, epochal change and transformative enhancement technologies. The synergistic convergence of nanotechnology, biotechnology, information technology and the cognitive sciences would seem to have little to do with the broad, almost metaphysical discussions of the future of humanity initiated in the previous section. Certainly there is some truth to this sentiment, and much in these reports is devoted precisely to scientific and technical discussions meant to foster ongoing efforts of convergence among scientists, researchers and engineers. And yet these reports are not limited to these matters in the least and in this section I will spell out some of their more radical elements connecting them to the concerns addressed in the previous section, focusing especially on Roco and Bainbridge's overview of NBIC convergence and their vision for converging technologies.

As we have seen, extending from Cassirer's 1930 essay 'Form and Technology' through his later work, especially *An Essay on Man* and *The Myth of the State*, is an awareness of a sense of crisis confronting the human being, a crisis brought on by, among other forces, the development of modern technology. A similar awareness is evident in Roco and Bainbridge's overview of NBIC convergence. And like Cassirer, this crisis has both a theoretical and a practical dimension. On the theoretical side, Roco and Bainbridge situate their discussion of converging technologies in a context that calls to mind Cassirer's own reflections on the fragmentation of knowledge. Like Cassirer, they are preoccupied with the fragmentation and specialization of human knowledge and the question of forging a general orientation, a frame of reference. They bemoan the specialization that has splintered the arts and engineering and worry that no one can master more than a tiny fragment of human creativity. In several passages that call to mind Cassirer's own interest in the Renaissance, Roco and Bainbridge suggest we need to return to a more holistic perspective and rekindle the spirit of the Renaissance. As they write in their overview to the report:

> The evolution of a hierarchical architecture for integrating natural and human sciences across many scales, dimensions, and data modalities will be required. Half a millennium ago, Renaissance artist-engineers like Leonardo da Vinci, Filippo Brunelleschi, and Benvenuto Cellini were masters of several fields simultaneously. Today, however, specialization has splintered the arts and engineering, and no one can master more than a tiny fragment of human creativity.[25]

Repeatedly, as well, the authors reference the holistic quality of the Renaissance as its hallmark: all fields of art, engineering, science and culture shared the same exciting spirit and many of the same intellectual principles.[26] With the development of science, that holism gave way to specialization and intellectual fragmentation, but Roco and Bainbridge argue that NBIC convergence will point the way to a new holistic spirit and a deeper level of unity based on the unity of nature.

Furthermore, Roco and Bainbridge, again in a move again suggestive of parallels to Cassirer, seem to intuit that this task is more than simply a theoretical need. We are, they write, 'caught in the grip of social, political, and economic conflicts, the world hovers between optimism and pessimism'.[27] And again they argue that NBIC convergence promises us relief, in part by providing the means to enhance human mental, physical and social abilities.[28] Bainbridge returns to this theme of crisis in a later essay on converging technologies, arguing that civilization is in such grave danger that we must seek a fresh foundation for our culture and institutions.[29] Bainbridge foresees a crisis brought on by an impending demographic catastrophe due to the collapse of fertility, the disintegration of families and social bonds in an aging global population, and the continued conflict between secular and religious societies. This crisis points to two possible futures for humanity. The first is a 'radical retrenchment that leads to a world fragmented among competing religious fundamentalisms'.[30] The second is made possible by converging technologies: 'A transcendence of the traditional human condition – made possible by the unification of all sciences and technologies, establishing a dynamic new creed to replace religion. This would open new worlds for humanity, not only in outer space, but also in the transformation of our own nature.'[31] NBIC convergence, then, becomes a new creed to replace religion, and provides the way forward that allows us to address the theoretical and practical crises facing the human being.

On the surface, we can see several intriguing similarities between Cassirer and Roco and Bainbridge, including an awareness of a sense of crisis, an understanding of the fragmentation of knowledge, the need for a fresh examination of our situation, and a call for holism and the unity of knowledge. But where Cassirer suggests that technology is implicated in this crisis and that what is needed is a fresh approach to the anthropological question, Roco and Bainbridge suggest that NBIC technologies offer a way out of this crisis, especially if they can converge on the improvement of human performance. Indeed, converging technologies and improving human performance provide the framework in which to address the various crises Roco and Bainbridge reference. Where Cassirer turned to a philosophical anthropology, Roco and Bainbridge turn to science and technology, especially nanotechnology, and a vision of convergence that begins with the material unity of all matter at the nano-scale. It is the convergence of the sciences that initiate the new renaissance and that convergence is based on material unity at the nano-scale and on technology integration from that scale. Roco and Bainbridge understand the natural world, human society and scientific research as 'closely coupled complex, hierarchical systems'.[32] They argue that we need to develop a hierarchical architecture for integrating the sciences beginning with physics as a base and moving up through chemistry and biology to psychology and economics.[33] They suggest we should not be concerned with the charge of reductionism and that all the sciences can progress through convergence: 'A trend towards unifying knowledge by combining natural sciences, social sciences, and humanities using cause-and-effect explanation has already begun and it should be reflected in the coherence of science and engineering trends and in the integration of R&D funding programs.'[34] Roco and Bainbridge's framework for convergence, then, is predicated on the material unity of nature, a hierarchical model of the disciplines founded on nanoscience, a holistic approach based on reductionistic and cause-and-effect models, all geared towards the improvement of human performance. More broadly, converging technologies serve as a model for other forms of convergence. Indeed, these authors suggest explicitly that we might once again achieve a golden age, a turning point, where 'technological convergence could become the framework for human convergence'.[35]

It is clear, too, that Roco and Bainbridge's proposal for NBIC convergence has profound implications for how we understand the human being. Indeed, the broad scope and implications of converging technologies are immediately suggested by the titles of the various reports edited by Roco and Bainbridge, including *Societal Implications of Nanoscience and Nanotechnology, Converging Technologies for Improving Human Performance, Converging Technologies in Society, The Coevolution of Human Potential and Converging Technologies*. The broad sweep of these reports is also indicated by the regular references to the 'immense individual, societal, and historical implications for human development'[36] and to the future of humanity, a future in which 'science and technology will increasingly dominate the world'.[37] Social progress, more than simply technological progress, is imagined to follow from the converging technologies. Bainbridge's 'Survey of NBIC Applications', included as an appendix to the report *Managing Nano-Bio-info-Cogno Innovations: Converging Technologies in Society*, tabulates 76 predicted applications of converging technologies, and includes, among other items, improvements to the human body, making it more durable, healthier, more energetic, easier to repair and more resistant to biological threats and aging; eradicating handicaps that have plagued the lives of millions of people; sociable technology that will enhance human emotional as well as cognitive performance, giving us more satisfactory relationships not only with our machines but also with each other;[38] devices connected directly to the nervous system that will significantly enhance human sensory, motor and cognitive performance.[39] These efforts at improving human performance are justified in light of the crises we face, crises motivated in part by the very technologies we are being encouraged to embrace.

Roco and Bainbridge suggest that the way out of the dilemma posed in the previous section is to embrace technology, improve the human being and transform society so that we are better able to meet the crises with which we are confronted. Their overview of NBIC convergence itself often converges on the utopian in this respect:

The twenty-first century could end in world peace, universal prosperity, and evolution to a higher level of compassion and accomplishment. It is hard to find the right metaphor to see a century into the future, but it may be that humanity would become like a single, distributed and interconnected 'brain' based in new core pathways of society. This will be an enhancement to the productivity and independence of individuals, giving them greater opportunities to achieve personal goals.[40]

Roco suggests that humanity will bond via this interconnected virtual brain of the Earth's communities as we search for intellectual comprehension and conquest of nature.[41] In 'Converging Technologies and Human Destiny' Bainbridge suggests that NBIC convergence will not only bring together science and technology but will also unite 'untraditional conceptions of reality with marvelously useful applications that cannot be ignored ... Once we use the technology to transform ourselves, then the technology becomes more salient for our hopes and beliefs than any ancient myth could be.'[42] Here too Bainbridge resorts to metaphor:

> Humanity is crossing an abyss on a tightrope. Behind us is the old world of religious faith that compensated wretched but fertile people for the misery in their lives. On the other side, if we can only reach it, is a new land where we no longer have to live by illusions, where wisdom and procreation are compatible, where truth and life are one. Nietzsche warned that as we make this perilous crossing, we must not look down.[43]

Converging technologies, Bainbridge suggests, may be that tightrope that can carry us to the other side.[44] Where once Cassirer urged a turn to philosophy of culture and a philosophical anthropology as a response to our crisis in self-knowledge, for Roco and Bainbridge that crisis calls for a focus on technology. Converging technologies, they suggest, provide a framework for reappropriating the holistic spirit of the Renaissance. A model for human convergence, NBIC technologies are the tightrope that will carry us beyond illusion and faith, to a world of enhanced human performance. What are we to make of this proposal? Does NBIC convergence provide a framework in which to address the dilemma of technology's impact on the human being? In order to assess this vision of converging technologies I would like to return once more to Cassirer and indicate his efforts at addressing this crisis and what elements we might find in it to critique Roco and Bainbridge's proposal. It is to that task that I turn in the next section.

III

In the midst of this very future-oriented, some might say utopian, discussion of the powers of nano-bio-info-cogno, turning back to Cassirer's

An Essay on Man as well as to an essay on technology written some 80 years ago might seem rather perverse. And yet, as we have seen, there are some interesting similarities, one might say convergences, between Cassirer's project in *An Essay on Man* and the project of NBIC convergence as suggested by Roco and Bainbridge. These similarities include a concern over crisis, the fragmentation of knowledge, the perceived need to bring some unity and cohesion to this fragmentation, inspiration derived from the Renaissance, and an interest in the human being as key to this 'convergence'. But here the similarities end and more interesting are the divergences between Cassirer and the proposed model of NBIC convergence. I maintain that in the search for a general orientation in which to address our sense of crisis, in discussions of converging technologies, changing societies and improving human performance, NBIC convergence, predicated on a unified understanding of the physical world from the nano-scale to the planetary scale, will prove inadequate to the task and is deeply problematic. More promising, I contend, is an alternative framework derived from a rich and multifaceted understanding of the human condition such as found in Cassirer's account of the 'animal symbolicum'. In order to support this claim, in this section I develop some of the main features of Cassirer's approach to philosophy of culture and philosophical anthropology, including his thoughts on technology.

An Essay on Man suggests that to face squarely man's crisis in self-knowledge we must arrive at a theory of man. But in order to develop a satisfactory account of the human being we need to understand the human being's particular milieu, culture, and in order to come to grips with human culture, we must engage in a study of the elements of culture, the symbolic forms. Our crisis in self-knowledge, then, impels Cassirer down a path wherein he must confront the multiplicity of symbolic forms: myth and religion, language, art, history, science and, importantly, technology. While Cassirer's philosophical anthropology, philosophy of culture and philosophy of symbolic forms are complex, multifaceted and spelled out over his lifetime, permit me to identify five key elements of which we should take particular note that are relevant to our discussion and which suggest Cassirer's unique position in the debate over technology and human enhancement.

First, Cassirer very explicitly situates the human being and culture in the organic realm. The work of the biologist Jakob von Uexküll and his account of the outward life and inward life of animals provides the backdrop to much of Cassirer's philosophical anthropology. Uexküll's study of animal form provides a way of avoiding the dualism of biology/culture life/spirit that Cassirer thought doomed previous philosophical anthropologies. As he puts it: 'A philosophical anthropology has to conform to the maxim of Spinoza that man is not to be regarded as a "state in the state." He is only a single link in the general chain of evolution. Cultural life is always bound up with the conditions of organic life.'[45] We must begin with the human being situated in his physical environment. The human being cannot live without constantly adapting himself to the conditions of the surrounding world.[46] Culture, the symbolic forms, do not represent the alienation of the human being from nature or an organic realm. Rather, the symbolic forms are the very conditions of human life.[47] And yet, while appropriating Uexküll's scheme, Cassirer argues that in the case of human beings, the functional circle between outward and inward life includes a new element, the symbol. Man, Cassirer observes, no longer lives in a merely physical universe but in a symbolic universe.[48] It is this qualitative change in human life that precludes Cassirer in 'Form and Technology' from judging technology according to a standard drawn from mere organic life.[49]

Second, the distinguishing feature of the human being is not some new feature or property, not some metaphysical essence. The human being's distinctiveness is his work. 'Man's outstanding characteristic,' Cassirer writes:

> his distinguishing mark, is not his metaphysical or physical nature – but his work. It is this work, it is the system of human activities, which defines and determines the circle of 'humanity'. Language, myth, religion, art, science, history are the constituents, the various sectors of this circle. A 'philosophy of man' would therefore be a philosophy which would give us insight into the fundamental structure of each of these human activities, and which at the same time would enable us to understand them as an organic whole.[50]

In focusing on this functional capacity of the human being, Cassirer avoids identifying human nature with some timeless metaphysical essence or substance.

Third, Cassirer insists on the diversity of the symbolic forms. His philosophy of symbolic forms represents a decisive break with the neo-Kantian tradition with which he is generally associated in recognizing that science was not the only manner in which human beings attempt to understand the world. Kant's Copernican revolution had to be extended to cover every principle by which human beings give form to the cultural world. In addition to investigating the function of cognition, then, we must also seek to understand the function of linguistic thinking, of mythical and religious thinking, artistic thinking. Each of these constitutes a symbolic form entirely independent of science with its own categories and concepts. Each is a particular way of seeing with its own measure and criterion of truth and meaning. As Cassirer notes: 'None of these forms can simply be reduced to, or derived from, the others; each of them designates a particular approach in which and through which it constitutes its own aspect of "reality".'[51] In regard to myth, for example, Cassirer is clear that while science aims at obliterating every trace of the mythic view, science cannot completely suppress myth. As he notes:

In the new light of science mythical perception has to fade away. But that does not mean that the data of our physiognomic experience as such are destroyed and annihilated. They have lost all objective or cosmological value, but their anthropological value persists. In our human world we cannot deny them and we cannot miss them; they maintain their place and their significance.[52]

The structure of *An Essay on Man* itself reflects Cassirer's commitment to the integrity and importance of each of the symbolic forms, with separate chapters on myth and religion, language, art, history and science. His goal in each of these chapters is similar to the goal he sets for himself in 'Form and Technology', to gain insight into the inherent, immanent law governing each form. Doing so establishes the heterogeneity of the forms of human culture.

Fourth, Cassirer situates his analysis of technology in the context of his philosophy of symbolic forms and in such a way that it would be inappropriate to conclude that technology represents the alienation of either culture or our nature as symbolic animals. In 'Form and Technology', it is clear that Cassirer wants to avoid overly quick and simplistic analyses of technology in terms of a blanket condemnation or praise of its effects or objects. 'We may bless technology or curse it, we may admire it as one of the greatest possessions of the age or lament its necessity and depravity – in judgments such as these, a measure is applied to it that does not originate from it.'[53] Philosophy's task is to enquire into the possibility of technology as a symbolic form, examining the form, meaning and essence of technology. And in this context, Cassirer draws very close parallels between technology and language as symbolic forms, emphasizing 'the affinity and internal connections that exist between technology and the pure form and principle of other basic powers of culture'.[54] Indeed, technology and language both emerge out of the magical-mythical worldview and represent two sides of the human essence. The human being is both a rational being and a tool-forming being. Both have their origins in the magical-mythical worldview, but where this worldview is defined by the immediacy of desire and subjective feelings of the wills, technology is defined by will and the growth in objectivity: 'in place of merely libidinous desire, there first emerges a genuine, conscious *wilfull* relationship'.[55] Language and tool use constitute a turning point for the human being, opening up a world of symbolic meaning.

Cassirer argues that we witness in technology a 'type of mediacy that belongs to the essence of thought'.[56] 'In its pure logical form, all thought is mediated. It is directed to the discovery and extraction of a mediating structure, which joins the opening sentence and the ending sentence of a communicative chain. The tool fulfils the same function, represented here in the logical sphere, in the objective sphere of physical objects.'[57] Language and tool use begin the human being on a slow and gradual process of growth – a progressive increase or strengthening of his self-consciousness. As Cassirer significantly notes, 'A new world-attitude and a new world-mood now announce themselves over and against the mythical-religious worldview. The human being now stands at that great turning point in his destiny and self-knowledge that Greek myth embodied in *Prometheus.*'[58] As we saw earlier in the first section of this essay, theorists such as Hayles and Bess suggest that with the development of new technologies we are witnessing a turning point in human life. For Cassirer, it is perhaps more correct to suggest that what makes possible the twenty-first-century references to a turning point lies far earlier in our history and with the development of the symbolic forms. The transition to the first tool, Cassirer suggests in 'Form and Technology', contains the turning point in knowledge, and that turning point comes in the opening up of the world of forms and culture and the break with the magical-mythic past, not in the particular developments of technology.

Cassirer's analysis of technology as a symbolic form precludes him from embracing the philosopher Ludwig Klages' account of technology as the alienation of human beings from their own essence. In fact, contrary to Klages' view, Cassirer, following philosopher of technology Ernst Kapp's suggestion, points out that knowledge of the 'I' is itself tied to the form of technical doing.[59] Agreeing with Kapp's basic perspective and insight, Cassirer notes: 'technological efficacy, when outwardly directed, likewise always exhibits a self-revelation and, through this, a means of self-knowledge'.[60] Cassirer emphasizes the conclusion that follows from this insight: 'with this first enjoyment of the fruit from the tree of knowledge the human being has cast himself out forever from the paradise of pure organic existence and life'.[61] Even when turning to modern, advanced technology, Cassirer continues to insist on the parallels between language and technology and draws the resulting inference that if we damn technology we must logically include in this condemnation 'the *totality* of culture'.[62] Cassirer makes a similar point in *An Essay on Man*, rejecting Rousseau's claim that it is a deterioration of human nature to exceed the boundaries of organic life: 'Yet there is no remedy against this reversal of the natural order. Man cannot escape from his own achievement. He cannot but adopt the conditions of his own life. No longer in a merely physical universe, man lives in a symbolic universe.'[63] Throughout his philosophical career, Cassirer remained critical of life philosophy with its emphasis on organicism and the immediacy of life and he rejected any blanket condemnation of culture according to some organic standard. This is not to suggest, though, that Cassirer was not critical of modern technology. But in order to grasp his critique of modern technology, I need to address a fifth and final element of his account of symbolic forms, their unity in terms of a common end.

A fifth important point we must make is to note that, for Cassirer, the diversity of symbolic forms does not preclude their unity. In his account of the symbolic forms, Cassirer emphasizes the perpetual strife of diverse conflicting forms. Philosophy, he cautions, cannot 'overlook the tensions and frictions, the strong contrasts and deep conflicts between the various powers of man'.[64] Each is a different step made by the human being in its reflective interpretation of life, an activity in which the human being attempts to make reality coherent, understandable and intelligible. And yet this multiplicity of forms does not, Cassirer says, denote discord or disharmony and it is precisely the task of philosophy to understand the system of culture as an organic whole. Philosophy begins with the hypothesis that the heterogeneous activities of human culture can be brought into a common focus.[65] But while Cassirer recognizes what we might think of as a kind of convergence among the symbolic forms, it is not a convergence that can be located in either nature or the metaphysical essence of the human being.[66] Nor can the diverse and heterogeneous forms of human culture simply be reduced to one form or placed into a fixed hierarchy.

The question of the unity of the symbolic forms is, in fact, central to Cassirer's understanding of the crisis of culture. 'Unless we succeed in finding a clue of Ariadne ... we can have no real insight into the general character of human culture; we shall remain lost in a mass of disconnected and disintegrated data which seem to lack all conceptual unity.'[67] As he makes this point in 'Form and Technology': 'It belongs to the essential task of philosophy to penetrate into this human law-giving, to gauge its unity and internal differences, its universality and differentiation.'[68] The various and conflicting symbolic forms are a coexistence of contraries held together in a dynamic and functional unity by a conformity in their fundamental task,[69] which Cassirer identifies in the final paragraph of *An Essay on Man* as the task of freedom: 'Human culture taken as a whole may be described as the process of man's progressive self-liberation. Language, art, religion, science are the various phases in this process. In all of them man discovers and proves a new power – the power to build up a world of his own, an "ideal" world.'[70] In both 'Form and Technology' and *An Essay on Man* Cassirer emphasizes how the world of culture, the symbolic forms, opens up to the human being a realm of freedom. In 'Form and Technology' Cassirer suggests that with the development of tools the human being is 'expelled onto a limitless path of creative work'.[71] Similarly, in some of the most evocative paragraphs of *An Essay on Man*, Cassirer describes how Helen Keller's grasp of the principle of symbolism is a magic key giving her access to the world of human culture. 'A new horizon is opened up, and henceforth the child will roam in this incomparably wider and freer area.'[72] We might say then that the symbolic forms converge on this task of freedom. It's clear, though, that this task is an ongoing one. There's a ceaseless struggle among the forms of human culture and we human beings have the task of bringing some equipoise to the centrifugal forces of human activity, in which especially the mythical elements are controlled by the constructive powers of logical and scientific thought, ethical forces and the creative energies of artistic imagination.[73]

Returning once more to Cassirer's discussion of technology, these final points indicate the direction of Cassirer's critique of modern technology. As we have seen, Cassirer rejects critiques of technology based on the claim that it alienates human beings from their organic life. As he notes, 'The standard by which it alone can be measured can, in the end, be none other than the standard of mind, not that of mere organic life. The law that one applies to it must be taken from the whole of the mental world of forms, not merely from the vital sphere.'[74] But in situating technology within the 'whole of the mental world of forms', Cassirer argues that it threatens to disrupt the equipoise of the coexistence of contrary symbolic forms. As he notes in 'Form and Technology':

> Moreover, as technology unfolds, neither does it simply place itself *next to* other fundamental mental orientations nor does it order itself harmoniously and peacefully with them. Insofar as it differentiates itself from them, it both separates itself from them and positions itself against them. It insists not only on its own norm, but also threatens to posit this norm as an absolute and to force it upon the other spheres. Here, a new conflict erupts within the sphere of mental activity, indeed, on its very lap. What is now demanded is no simple confrontation with 'nature', but the erection of a barrier within mental life itself – a universal norm that both satisfies and restrains individual norms.[75]

While Cassirer observes that the human being is 'thrown by technological culture into a never-ending vertigo that moves from desire to consumption, from consumption to desire',[76] he argues that this is not due to the form of technology but to its connection with a 'certain *form and order of commerce*', a concrete historical position.[77] The more basic problem is technology's setting itself up as a leader and an end itself rather than a servant collaborating to carry out goals in the context of the ethical task of culture as man's progressive self-liberation. The danger that technology presents is that it usurps the other symbolic forms, the unity of the symbolic forms, and sets itself up as the dominant if not sole symbolic form. Cassirer's analysis of technology points the way towards a critique of Roco and Bainbridge's proposal for NBIC convergence, to which I turn in the next section.

IV

In the previous three sections of this essay, I have drawn parallels between Cassirer's and Roco and Bainbridge's understanding of the crisis we human beings face in self-knowledge, owing in part to the advance of technology. Furthermore, I have sketched out core elements in their respective frameworks proposed to address this crisis, examining the main elements of Roco and Bainbridge's vision for converging technologies focused on improving human performance, and Cassirer's account of a philosophical anthropology focused on the human being as a symbolic animal. In this final section of the essay I would like to bring the two strands of this conversation into closer contact and – from the perspective of Cassirer's account of a philosophy of technology, a philosophy of symbolic forms and his philosophical anthropology – argue that Roco and Bainbridge's framework of NBIC convergence and the focus on improving human performance will prove inadequate to the task of confronting the stark issues we face as we contemplate the development of our technics and powers. Having spelled out some of the important elements of Cassirer's account of a philosophy of culture, including his account of the human being, the symbolic forms and technology, let us return to the epochal shift, the turning point, we are facing, given the emergence of converging technologies and their impact on improving human performance.

Focusing on the broad elements of their proposed frameworks, the differences between Roco and Bainbridge's and Cassirer's frameworks could not be starker. For Roco and Bainbridge, the focus remains almost exclusively on science and technology and the impetus largely has to do with managing a crisis rather than examining its underlying causes. They suggest that if we can put in place the appropriate bureaucratic and scientific structure we can forge a convergence of all disciplines around a unified set of causal principles that will powerfully give rise to a transformative science and technology that will in turn have radical implications for human society and the human condition. While occasionally incorporating the humanities into their account of a more holistic convergence and observing that proper attention must be paid to ethical issues and societal need, Roco and Bainbridge's vision focuses on NBIC convergence, with technological convergence providing the framework for the unity of disciplines, predicated on nanotechnology, the unity of nature, cause-and-effect thinking and a hierarchy of disciplines. Furthermore, their model of convergence is built on the assumption that reductionism will lead to the unity of the arts and humanities with the scientific disciplines.

Like Roco and Bainbridge, Cassirer too is ultimately interested in a kind of convergence. But his deep and abiding interest in the forms of human culture convinced him that this unity cannot simply be read from nature. There is only one world and we are a part of it but our efforts to understand and make sense of ourselves as part of this world are plural and divergent and resist the kind of linear, hierarchical and ultimately reductive model proposed in many of the converging technologies proposals. Ultimately, the notion that technological convergence could become a model for human convergence is still held hostage to the mechanistic and machine metaphysic of the seventeenth century that Cassirer, in the best tradition of philosophical anthropology, attempts to move beyond. Science is one of the activities of the human being, but it is not the only one and, as a symbolic form, neither science nor technology can serve as a model for convergence in the humanities or in human culture more broadly. Cassirer argues that in order to come to grips with the crisis of the fragmentation of human knowledge we need a philosophy of human culture and he offers a rich analysis of the various symbolic forms that are constitutive of human culture. 'Every feature of our human experience', Cassirer argues, 'has a claim to reality'[78] and each symbolic form has its value. 'None of them is a mere illusion; every one is, in its measure, a step on our way to reality.'[79] Throughout *An Essay on Man*, Cassirer is critical of hypostasizing scientific models and principles as simply mirroring nature and providing unmediated access to reality.

Cassirer's analysis of the form of technology also suggests that he would be sceptical of looking to technology for a model of convergence. Behind Roco and Bainbridge's proposal for NBIC convergence focused on improving human performance is a recognition that the rapid progress of science and technology threatens to outstrip human capabilities. Their solution to that dilemma is to employ the very same science and technology to transform the human being and create a new and improved man better able to function in a society shaped by NBIC convergence. Cassirer's analysis of technology demonstrates that he is no technophobe and he repeatedly rejects the false dichotomy of alternatively praising or blaming technology. He does caution, though, that the danger of modern technology is that it threatens to posit itself as an absolute norm relative to the other forms of human culture. It is in this context that Cassirer raises the charge voiced by the writer, industrialist and politician Walter Rathenau that modern technology is the water jug of the Danaides, observing that the human being is thrown by technological culture into a never-ending vertigo.[80] Reading the proposals for NBIC convergence, one can be struck by the perception that technological development and the improvement of the human being become locked in a tight circle of constant innovation and development. A cycle from which, as Cassirer observes, there is no escape. The human being is to be improved by and for technology.

The difficulty, Cassirer argues, is that technology sets itself up as an absolute norm and an ethical task is transformed into a technical task to be managed. And yet, Cassirer insists, technology cannot be a leader here but a servant. We must bring to technology the ethical question, not draw our ethical concerns from technology. You cannot derive an ethics from within the culture of technology and the problems of technology cannot be undone by means of technology alone. Cassirer forces us to situate technology within the realm of the other symbolic forms and within the context of the task of the symbolic forms in terms of the progressive self-liberation of the human being. For Cassirer, progress, far from being a scientific task, is an ethical task and a perpetual one. As he notes in *An Essay on Man*, the ethical world is never given; it is forever in the making. Cassirer identifies in human culture a fundamental polarity between innovation and stabilization. 'Man', he writes, 'is torn between these two tendencies, one of which seeks to preserve old forms whereas the other strives to produce new ones.'[81] Our task is to struggle to bring these forces into some equilibrium. There is equipoise to maintain here that cannot be sought from any one of the cultural forms but must be considered an ongoing dynamic task of the human being in culture. Taking all of our cues from technology and technical innovation will surely upset this equipoise.

Cassirer concludes 'Form and Technology' by suggesting that technology best understands its own meaning and narrative when it is 'content in the fact that it can never be an end itself'.[82] One goal of his essay is to understand technology within human culture and the other symbolic forms and caution us against creating ethical values out of technology itself. Technical enhancements are not, Cassirer might argue, the fundamental human enhancement. Rather, that is to be found in the spontaneous and productive construction of the symbolic universe[83] and efforts to foreclose upon that capacity are at odds with our symbolic nature. The converging technologies reports do precisely that by virtue of their narrow, hierarchical and reductive approach to convergence. In their single-minded focus on NBIC technologies and their efforts to realign human knowledge on a foundation of the nanosciences and the building blocks of matter, Roco and Bainbridge's proposal for convergence work to marginalize human culture and transform our symbolic behaviour into calculative behaviour.

The failure of their proposal for convergence is apparent in their own symbolic constructs, including their references to global brains, tightropes and other metaphors. While Bainbridge suggests that converging technologies moves us beyond myth and illusion, and Roco and Bainbridge argue for a model of convergence predicated on science and technology, we have seen that at key moments in their exposition they fall back on precisely the myths and metaphors their framework seemingly excludes. Here we are in the realm of what Cassirer, in a different context, refers to as rationalized myth:

the blending together and even complete fusion of two contradictory and incompatible elements: of the elements of magical thinking and technical thought. The modern politician had to combine in himself two entirely different functions. He had to be a homo magus and a homo faber at the same time. He was the spokesman for and the priest of a new and entirely irrational and mysterious religion.[84]

Converging technologies becomes a new secular myth, meant to replace outmoded religious myths no longer up to the challenge of facing the crisis we human beings face.

It is here that we witness the ultimate failure of convergence, the emergence of myth in the realm of science and technology. In Roco and Bainbridge's recognition of a crisis in thought and in their efforts to forge a convergence of NBIC technologies, we see how myth cannot be denied. Man, Cassirer notes, is not exclusively a rational animal; he is and remains a mythical animal. Myth is part and parcel of human nature.[85] Roco and Bainbridge would have us transcend the human condition in our embrace of converging technologies and yet their proposed framework fails to account for key elements of its own proposed vision. While NBIC convergence is predicated on technology, cause-and-effect thinking and the material unity of nature, at key moments in its defence we see emerge elements of myth and metaphor, elements Cassirer was clear cannot be eradicated from human culture: 'there is no danger that mankind ever will forget or renounce the language of myth. For this language is not restricted to a special field; it pervades the whole of man's life and existence.'[86] Science too, no less than language and art, is a symbolic form, employs metaphor and is intimately connected to mythic thought. Myth, Cassirer suggests, is the common background and common basis for all the symbolic forms and while science attempts to extirpate its mythic roots it can never free itself entirely from myth and metaphor.

Cassirer suggests that science offers us a freer and larger horizon of knowledge.[87] Science affords us the self-critical awareness of its mythic roots. But science too can turn dogmatic and Cassirer warns against hypostatizing the categories of science as the ultimate reality. As we have seen, Cassirer cautions against taking any of the forms, whether myth or science or technology, as absolute norms. This is precisely what happens in times of crisis and it is at moments of crisis that we are most susceptible to the power of myth. In such times, we are prone to distrust our formative and creative powers and fall back to the mythic. As Cassirer notes in his final work, *The Myth of the State*:

> In all critical moments of man's social life, the rational forces that resist the rise of the old mythical conceptions are no longer sure of themselves. In these moments the time for myth has come again. For myth has not been really vanquished and subjugated. It is always there, lurking in the dark and waiting for its hour and opportunity. This hour comes as soon as the other binding forces of man's social life, for one reason or another, lose their strength and are no longer able to combat the demonic mythical power.[88]

When man is confronted with a task that seems to be far beyond his natural powers, he returns to the realm of myth and magic. And it is here that we see the breakdown in the bureaucratic rationale for NBIC convergence. Where so much of the report is geared towards productivity, efficiency, promoting better lines of communication among scientific and technical disciplines, in the cracks of that bureaucratic scheme emerge myth and metaphor. And it is here that we most witness the failure of convergence. When science and technology turn dogmatic, when they abrogate to themselves the sole power to address the crises facing human beings, they outstrip their function as symbolic forms. For Cassirer, convergence cannot simply be read from nature, nor can it come simply as a matter of coordinating a few specialized scientific and technical disciplines. Technology is central to human life, as Cassirer makes clear in 'Form and Technology'. But he also makes clear that technology cannot set itself up as an absolute norm forced upon the other symbolic forms. Technology, he reminds us, cannot determine the goal, though it should collaborate in carrying it out.[89] As one of a number of symbolic forms, technology has a role to play in the human being's ethical task of balancing the tensions and frictions, contrasts and conflicts between the symbolic powers of the human being.[90] Cassirer's notion of the 'animal symbolicum' presents us with a rich and multifaceted philosophical anthropology that ultimately frustrates neater and simpler efforts at convergence. The task of coming to terms with our technics and powers and addressing the crises we face is a human crisis, not merely a technical crisis, and its resolution comes in our continued efforts to grapple with our symbolic nature.

Notes

1. Ernst Cassirer, *An Essay on Man* (New Haven: Yale University Press, 1944), 22.
2. *Ibid.*, 21–2.
3. *Ibid.*, 22.
4. Ernst Cassirer, *Symbol, Myth, and Culture*, ed. Donald Phillip Verene (New Haven: Yale University Press, 1979), 60.
5. *Ibid.*, 233.
6. Arnold Gehlen, *Man: His Nature and Place in the World*, trans. Claire McMillian and Karl Pillemer (New York: Columbia University Press, 1988), 4.
7. Michael Landmann, *Philosophical Anthropology*, trans. David Parent (Philadelphia: Westminster, 1974), 6.
8. Paul Pfuetze, *The Social Self* (New York: Bookman Associates, 1954), 19.
9. See Ernst Cassirer's 'Form and Technology' in this volume, 16.
10. *Ibid.*, 35.
11. *Ibid.*, 48.
12. Susan Squier, *Liminal Lives* (Raleigh: Duke University Press, 2004), 2.
13. Katherine Hayles, 'Preparing the Humanities for the Posthuman', *News of the National Humanities Center*, Autumn/Winter (2007), 1–14.
14. Michael Bess, 'Icarus 2.0', *Technology and Culture*, 49 (2008), 114.
15. Joel Garreau, *Radical Evolution: The Promise and Peril of Enhancing our Minds, Our Bodies – and What it Means to be Human* (New York: Doubleday, 2005), 3.
16. Garreau, *Radical Evolution*, 21.
17. Bill Joy, 'Why the Future Doesn't Need Us', *Wired*, April (2000), 238.
18. Ray Kurzweil, *The Age of Spiritual Machines* (New York: Viking, 1999), 2.
19. Ray Kurzweil, *The Singularity is Near: When Humans Transcend Biology* (New York: Viking, 2005), 90.
20. *Ibid.*, 23.
21. *Ibid.*, 7.
22. *Ibid.*, 136.
23. *Ibid.*, 415.
24. See Mihail C. Roco and William Sims Bainbridge (eds), *Converging Technologies for Improving Human Performance* (Dordrecht: Kluwer, 2003); Mihail C. Roco and Carlo Montemagno (eds), *The Coevolution of Human Potential and Converging Technologies* (New York: New York Academy of Sciences, 2004); and W. S. Bainbridge and M. Roco (eds), *Managing Nano-bio-info-cogno Innovations: Converging Technologies in Society* (Dordrecht: Springer, 2006).
25. Roco and Bainbridge, *Converging Technologies for Improving Human Performance*, 13.
26. *Ibid.*, 4.
27. *Ibid.*, 3.
28. *Ibid.*
29. *Ibid.*, 197.
30. *Ibid.*, 198.
31. *Ibid.*
32. *Ibid.*, ix.
33. *Ibid.*, 13.
34. *Ibid.*
35. *Ibid.*, 6.

36 *Ibid.*, xiii.
37 *Ibid.*
38 William Sims Bainbridge, 'Survey of NBIC Applications. Appendix 1', in *Managing Nano-bio-info- cogno Innovations*, 342.
39. *Ibid.*
40. *Ibid.*, 6.
41. M. C. Roco, 'Scientific and Engineering Innovation in the World: A New Beginning', SATW, 23 September 1999, Zürich – Aula der Eidgenössischen Technischen Hochschule Switzerland.
42. W. S. Bainbridge, 'Converging Technologies and Human Destiny', *Journal of Medicine and Philosophy*, 32:3 (2007), 202.
43. Bainbridge, 'Converging Technologies and Human Destiny', 202–3.
44. *Ibid.*, 209.
45. Cassirer, *Symbol, Myth, and Culture*, 168.
46. Cassirer, *An Essay on Man*, 3.
47. *Ibid.*, 25.
48. *Ibid.*
49. Cassirer, 'Form and Technology', 41.
50. Cassirer, *An Essay on Man*, 68.
51. Ernst Cassirer, *The Philosophy of Symbolic Forms*, vol. 1: *Language*, trans. Ralph Manheim (New Haven: Yale University Press, 1971), 78.
52. Cassirer, *An Essay on Man*, 77.
53. Cassirer, 'Form and Technology', 21.
54. *Ibid.*, 22.
55. *Ibid.*, 30.
56. *Ibid.*
57. *Ibid.*
58. *Ibid.*, 33–4.
59. *Ibid.*, 37–8.
60. *Ibid.*, 38.
61. *Ibid.*
62. *Ibid.*, 41.
63. Cassirer, *An Essay on Man*, 25.
64. *Ibid.*, 228.
65. *Ibid.*, 222.
66. *Ibid.*
67. *Ibid.*, 22.
68. Cassirer, 'Form and Technology', 22.
69. Cassirer, *An Essay on Man*, 222.
70. *Ibid.*, 228
71. Cassirer, 'Form and Technology', 34.
72. Cassirer, *An Essay on Man*, 35.
73. Cassirer, *Symbol, Myth, and Culture*, 246.
74. Cassirer, 'Form and Technology', 41.
75. *Ibid.*, 42.
76. *Ibid.*, 48.
77. *Ibid.*, 49.
78. Cassirer, *An Essay on Man*, 77.
79. *Ibid.*, 78.

80. Cassirer, 'Form and Technology', 48.
81. Cassirer, *An Essay on Man*, 224.
82. Cassirer, 'Form and Technology', 49.
83. Cassirer, *An Essay on Man*, 221.
84. Cassirer, *Symbol, Myth, and Culture*, 253.
85. *Ibid.*, 246.
86. *Ibid.*, 245.
87. Cassirer, *An Essay on Man*, 77.
88. Ernst Cassirer, *The Myth of the State* (New Haven: Yale University Press, 1946), 280.
89. Cassirer, 'Form and Technology', 49.
90. Cassirer, *An Essay on Man*, 228.

Index

Abenteuerliche Herz, Das (Ernst Jünger), 93, 96
Adorno, Theodor W., 133
Aesthetic perception, 161–5, 189–94
  connection to consciousness, 194
  insularity of, 193
  suddenness of, 193
'Age of the World Picture, The' (Martin Heidegger), 8, 114–15, 118, 125, 129, 130–2
Alberti, Leon Battista, 45
Alienation, 35, 42, 49, 73, 82–3, 102, 107, 119, 122, 123, 124, 164, 207, 210, 244–6
Alienation effects, 210
  see also Distancing strategies, Verfremdungs-Effekt,
Arbeit und Rhythmus (Karl Bücher), 39
Arbeiter: Herrschaft und Gestalt, Der (Ernst Jünger), 92–108
Arendt, Hannah, 96
Aristotle, 4, 54, 74–5, 77, 79, 86, 151, 218
  see also Poetics (poiesis), Practice (praxis), Theory (theoria),
Art, 3, 46, 54, 163, 164–5, 189, 190, 191, 214–32
  cave art, 214–32
  contribution to social change, 164–5
  as creator of humanity, 222
  participation in, 163
  presence and representation in, 189
  rock art, 215, 217
  status of artists in Plato, 54
  symbolic form of, 191
  technological and artistic creation compared, 46
  see also Creation, Formation, Symbolic forms, Work
Artemis, 148

Auseinandersetzung, 5, 17–18, 26, 30, 36, 115, 172–3, 220
  as mutual differentiation and determination, 26, 30, 36, 173
  productive act of determining through processes of differentiation, 5, 173, 220
  genuine polarity, 17
  see also Difference, Formation, Mediation, Work

Bacon, Francis, 87
Bal, Mieke, 163–4, 170, 172
Balzac, Honoré de, 209
Basis phenomena, 68, 72, 78, 79
Beauty, 21, 36, 45–7, 107, 125, 226
Becoming, 18, 22, 78, 221, 228
  see also Concept of being
Being and Time (Martin Heidegger), 101, 114, 115, 127
Being of technology, 22, 65, 67, 105
  becomes visible only in becoming, 22
  grasped in activity, 22
  technologies ensembles of vector-magnitudes, 32
  see also Becoming, Concept of being, Essence of technology, Question of being, 'Question Concerning Technology, The'
Berlin: Die Sinfonie der Großstadt (Walter Ruttmann), 59
Bergson, Henri, 92
Biology, 147, 237, 240, 244
Boeckmann, Kurt von, 168
Bohrer, Karl-Heinz, 193
Brecht, Bertolt, 8, 161–78
Broch, Hermann, 209
Bücher, Karl, 39
Bühler, Karl, 142

Camus, Albert, 207
Capitalism, 96, 105
Carnap, Rudolf, 113

Cassirer, Hugo, 56
Causality, 10, 26–7, 31–2, 121, 192,
    240, 250
    as habitual connection (Hume), 31
    as necessary connection (Kant), 31
Cave art, 10, 214–32
Celan, Paul, 190
Clark, Andy, 85
    see also Extended mind
Clegg, John, 215
Clottes, Jean, 216–17
Coetzee, John Maxwell, 207
Cognition, 66, 68–71, 76, 84–5, 87,
    107, 147, 149, 203, 205, 235,
    245, 251
    pluralization of cognitive function,
    70
    tools of the mind, 3, 23, 39, 71, 83
Cognitive forms, 9, 200
    see also Symbolic forms
Cognitive science, 135
Cohen, Hermann, 77–8, 114
Coleman, James, 163, 172
Concept of being, 19–20, 100
    versus concept of form, 19–20
Concept of form, 19–20, 46, 100
    versus concept of being, 19–20
Consciousness
    moral, 49
    representational character of, 186
    sensory, 186, 187
Constructive technology assessment,
    83
Constructivist theory of
    representation 9, 182, 189, 217,
    218
Copy theory, 43, 68, 182, 188, 189
Correspondence, 43–4, 65, 74, 80–1,
    184
    between thought and reality, 43
    genuine *adaequatio*, 44, 81
Creation, 9, 18, 24, 32, 35, 36, 41,
    45–6, 102, 121, 124,127, 129,
    183, 208–13, 216, 228, 242
    artistic creation, 45–6
    versus exposition and transposition,
    208–13
    of the human being, 36, 102
    of the human mind, 33, 35

of the world as image, 127
world-creation, 24, 129
    see also Formation, Mediation,
    Technological creation, Work
Cristuado, Wayne, 133
Critical idealism, 135
Critical theory, 10, 162, 165, 176
Critique of Judgment (Immanuel Kant),
    21, 79
Critique of Pure Reason (Immanuel
    Kant), 68, 70, 86, 116, 203
Cybernetics, 135

Danzel, Theodor-Wilhelm, 4, 25, 133
Davos meeting, 1–2, 7–8, 11, 77–8,
    92, 113–17, 133–5
Deconstruction, 9, 189
Democracy, 8–9, 94, 98, 102, 161–2,
    164, 168–9, 170
Derrida, Jacques, 5, 9, 188, 190,
    206–7
    see also Deconstruction, Difference,
    Poststructuralism
Descartes, René, 84, 126–7
Dessauer, Friedrich, 19, 100, 119,
    124
Destiny (Schicksal), 7, 8, 20, 34, 59,
    92, 93, 94, 98, 105, 106, 113–38
Dialectics of Enlightenment
    (Theodor W. Adorno and Max
    Horkheimer), 133
Difference
    deconstructive difference theories,
    189
    differential approach to knowledge,
    7, 66–7, 69
    differential sight, 81
    positive notion of, 3
    process of differentiation, 5
    separation of I and reality, 26
    see also Auseinandersetzung,
    Formation, Mediation, Work
Distancing strategies, 177
    see also Alienation effects,
    Verfremdungs-Effekt
Dostoyevsky, Fyodor, 207
Dreyfus, Hubert, 88
Dunlavey, Wilson McClelland, 6,
    194

Economy, 16, 48–9, 96, 97
of literature, 206
and technology, 48–9
see also Capitalism
Efficacy, 15
Eiffel Tower, 47
*Eighteenth Brumaire of Louis Bonaparte,
The* (Karl Marx), 204
Einstein, Alfred, 56–7
Emotions, 46, 172, 191, 193, 208,
241
manipulation of, 59
see also Expression, Imagination
'En-framing, The' (Martin Heidegger),
92, 130
Enlightenment, 7, 21, 92, 94, 95, 101,
107, 108, 170
Ephesus, 148
Essay on Man, An (Ernst Cassirer),
5, 233–7, 238, 243, 245, 247–8,
251–2
Essence of technology, 17, 19, 22, 67,
105, 123, 130–4, 245
see also Being of technology,
'Question Concerning
Technology, The'
Ethics, 8, 10, 20, 21, 46–50
ethicization of technology, 49
Ethology, 8, 27, 144
Euclidean geometry, 218
Experimental work, 10, 83, 88, 124,
126, 157, 161, 170, 171, 178, 184,
207, 208, 219–33
in aesthetic practices,161, 170, 171,
178, 207, 208
in cave art research, 10, 219–33
scientific relevance of, 83, 88, 124,
126, 157, 184, 226, 233
Exposition, 10, 173, 208–13
versus creation and transposition,
208–13
Expression, 46, 140, 162, 172, 173,
178, 182, 187, 192, 193
expressive forms of reason, 193
expressive function, 172, 182
expressive meaning, 178, 192
expressivity of language, 140
see also Myth, Perceptive activity,
Symbolic pregnance

Extended mind, 7, 84, 85, 99
Eyth, Max, 19, 22–3, 71, 100
mental aspect of the physical tool,
23
tools of the mind, 23
word and tool, 22–3

Fascism, 98, 107
Fiction, 80, 155, 164, 185, 200, 205,
206, 207, 208ff
Flaubert, Gustave, 209
Flight of the Lindberghs, The (Bertolt
Brecht), 170
Flying techniques, 39, 54, 103
Folkvord, Ingvild, 8, 161
Formation, 4, 15, 16, 18, 22, 23, 24,
26, 28, 37, 45, 48, 68, 70, 72, 74,
82, 87–8, 96, 122, 162, 166, 169,
172, 176, 182, 183, 186–7, 221,
254
double act of grasping, 24
form of the world not received but
built, 24
forma formata, 18
forma formans, 18
formative energy, 15
formative impulse, 22
formative intervention, 74
formative power (capacities,
forces), 68, 70, 87–8, 176, 182,
254
formative process, 18, 72, 221
formative tool, 162, 166, 169
principle of formation of
technology, 48
symbolic formation, 183
world-formation, 28, 166
see also Auseinandersetzung,
Difference, Mediation,
Technological formation,
Transformation, Work
Foucault, Michel, 133, 181, 201, 217
Frankfurt School, 6
Franklin, Benjamin, 24
tool-making animal, 24, 66, 80
Frazer, James George, 26, 27, 31, 100,
101, 121
Free possibilities, 44, 81, 131
and necessity, 44

Freedom, 2, 21, 34, 41, 47, 49, 99,
    101, 102, 104, 106, 107, 108,
    123–4, 128, 132, 248
Freudenthal, Gideon, 4
Friedman, Michael, 113
Functionalism, 84–5

Galilei, Galileo, 43
GAS (Georg Kaiser), 58
German idealism, 1, 21, 117, 121,
    127, 129
    transcendental philosophy, 17, 68
    see also Immanuel Kant
'Ge-stell, Das' (Martin Heidegger), 92,
    130
Gideon, Siegfried, 54–5
Goethe, Johann Wolfgang von, 1, 22,
    33, 37, 42, 103, 117, 166, 185–6,
    192
Goodman, Nelson, 190
Gordon, Peter, 113
Gorgias (Plato), 48
*Grundlinien einer Philosophie der
    Technik* (Ernst Kapp), 37, 38, 119,
    220
Gumbrecht, Hans Ulrich, 9, 188, 189,
    193

Habermas, Jürgen, 133
Hacking, Ian, 88
Hardt, Ernst, 168
Havelock, Eric A., 152
Hedonism, 35, 48, 105, 106, 107
Hegel, Georg Wilhelm Friedrich, 16,
    121, 122, 141, 181
Heidegger, Martin, 1, 6, 7,8,11, 69,
    77, 78, 92, 93, 96, 101, 113–38,
    176
    see also Davos
Henningsen, Bernd, xi
Heraclitus, 187
Herder, Johann Gottfried, 33, 36, 140
Hermeneutics, 4, 115, 162, 165, 166,
    188, 192, 215
Herodotus, 148
Herrenschmidt, Clarisse, 148
Hitler, Adolf, 59
Hoel, Aud Sissel, 7, 65
Hogarth, David George, 148

Holism, 101, 239, 240
Holm, Isak Winkel, 9, 199
Homo divinans, 25, 28, 121
    versus homo faber, 25
Homo faber, 25, 121, 253
    versus homo divinans, 25
Horkheimer, Max, 133
Humboldt, Wilhelm von, 4, 22, 24,
    36, 142, 166
Husserl, Edmund, 69, 89, 99, 113,
    114–16, 131, 132

'Ideal and Life, The' (Johann,
    Christoph Friedrich von Schiller),
    175
Ihde, Don, 82–3, 88
    postphenomenology, 82
Imagination, 28–9, 76, 78, 79, 88,
    116, 162, 172, 187, 203, 205, 236,
    248
    see also Creation, Expression,
    Formation, Mediation

Jonny spielt auf (Ernst Krenek), 59
Joulian, Frédéric, 144
Jünger, Ernst, 7, 92, 93–109, 106, 107,
    120, 125

Kant, Immanuel, 8, 9, 17, 43, 49,
    66–71, 75, 76, 77, 79, 86, 87, 101,
    115–18, 120, 121, 123, 127, 128,
    129, 132, 135, 152, 166, 182, 185,
    186, 189, 190, 193, 202, 203, 205,
    218, 243
    modern science and experience,
    152
    realm of the beautiful, 21
    transcendental philosophy, 17,
    121
Kant and the Problem of Metaphysics
    (Martin Heidegger), 113, 115
Kapp, Ernst, 37–8, 103, 119, 122, 147,
    148, 149, 154, 220, 222, 246
    see also Organ-projection
Kestenberg, Leo, 7, 55–6, 59, 167, 169
Kierkegaard, Søren, 117
Kittler, Friedrich, 165
Klages, Ludwig, 35, 73, 92, 102, 105,
    122, 235, 246

Knowledge, 7, 16, 17, 18, 23, 26, 27, 30, 38, 42, 46, 47, 65, 66–7, 68, 69, 70, 71, 72, 74, 75, 76, 77, 80, 82, 86, 88, 96, 116, 126, 127, 134, 141, 146, 173, 174, 176, 184, 185, 199–202, 203, 205, 214, 218, 233, 238, 239, 240, 243, 246, 247, 251–2, 253
  versus action, 16
  beauty of, 46
  critical theory (critique) of, 69, 116, 184, 203
  cultural knowledge, 201
  differential approach to, 7, 66–7, 69
  empirical knowledge, 218
  fragmentation of, 238, 240, 243
  genetic view of, 68
  human knowledge, 68, 82, 86, 233, 239, 251–2
  knowledge of subject and object, 26, 173
  life complex becomes knowledge complex, 72
  made possible by technology, 65
  objectivity of, 80, 88
  problem of, 7, 74
  process of, 86
  productive view of, 71
  rational knowledge, 205
  reframing knowledge, 75
  scientific knowledge, 27, 127, 184
  spectator theory of, 75
  system of, 127
  theoretical knowledge, 17, 18, 42, 44, 47, 70, 77, 80
  theory of, 16, 18, 69, 71, 75, 116, 214
  tool-use as turning point in knowledge, 30, 76, 246
  tree of knowledge, 38, 247
  truth and facts, 199–202
  unity of, 174, 240
  see also Cognition, Formation, Logos, Mediation, Self-knowledge, Thinking
Krenek, Ernst, 59
Krois, John Michael, v, xi, 6, 50, 54, 101, 106, 167

Kunst und Technik (Leo Kestenberg), 55, 167–9

Laming-Emperaire, Annette, 228
Lang, Fritz, 59
Langer, Susanne, 191
Language, 2, 3, 4, 5, 8, 18, 22–5, 26, 33, 36, 39, 70, 71, 80, 85, 86, 87, 92, 99, 104, 130, 139, 140–3, 145–7, 151–7 155, 165, 166, 167, 169, 172, 173, 174, 177, 183, 188, 189, 207, 208, 209, 210, 218, 221, 222, 227, 243, 244, 245, 245, 246, 247, 248, 253
  as creator of humanity, 36
  expressivity of, 140
  genetic definition of, 22
  human attitude towards natural language, 139
  modifying function between symbolic forms, 140
  presentative phenomena in, 188
  relationship to money, 151–7
  relationship to technical activity, 22–5, 140–3, 166, 221
  relationship to tool-making, 22–5, 145–7
  see also Herder, Humboldt, Symbolic forms
Lassègue, Jean, 8, 139
Latour, Bruno, 82, 83
Lauschke, Marion, 9, 181
Lebensphilosophie, 16, 92
Leibniz, Gottfried Wilhelm, 44, 183–4, 192
  theory of monads, 183
  see also Monads
Lenay, Charles, 150
Leonardo da Vinci, 43, 45, 75, 150, 151, 239
Lerol-Gourhan, André, 228
Listening reception, 56, 132, 161–78
  intimacy of, 162
  passivity of, 163, 165
  see also Brecht, McLuhan, Radio
Literature, 9, 58, 72, 190, 199–213
  literary deviation, 9, 211–13
  literary interventions in political affairs, 210–11

Literature – *continued*
  literary representational practices,
    190, 199–213
  the politics of, 206
  see also Creation, Formation, Poetics
    (poiesis), Symbolic forms
Logical empiricism, 57
Logical Investigations (Edmund
    Husserl), 114, 115
Logos, 23, 70–1
  expansion and transformation of,
    70–1
  instrumental and theoretical
    meaning of, 23
Lorblanchet, Michel, xii, 10, 219,
    221–7
Lukács, Georg, 209

Magic Art, The (James George Frazer),
    26
Magical desire, 25
  versus technological will, 25
Magical-mythical worldview, 25–34
  compared with technological
    efficacy and scientific thinking,
    28–34
Man Equals Man (Bertolt Brecht), 174,
    176
Marx, Karl, 8, 38, 96, 106, 119, 147,
    162, 174, 204
  emancipation of the organic barrier,
    38
Maschinist Hopkins (Max Brand), 58, 59
Mauss, Marcel, 144, 145
McLuhan, Marshall, 162–5, 170,
    171
Mechanization, the age of complete
    mechanization, 55, 56, 59, 60
Mechanization Takes Command
    (Siegfried Gideon), 55
Mediation, 2–3, 7, 30, 65, 72–4, 82,
    83, 101, 104, 140–1, 149, 153,
    154, 161–78, 246
  and difference, 72–4
  as dynamic, double and two-way
    process, 2, 153, 169
  intervening medium, 3, 83, 101
  as involving productive distance,
    2, 162–6

linguistic mediation, 140–1, 149, 154
  mediating process, 3
  radiophone mediation, 8, 161–78
  technology as primary medium, 3
  terminus medius, 101
  and thinking, 7, 30, 65, 101, 246
  transformation, 3
  see also Auseinandersetzung,
    Difference, Formation,
    Technology, Tools,
    Transformation, Work
Mensch und die Technik, Der (Oswald
    Spengler), 119
Mendel, Gregor, 201
Mesopotamia, 148
Metaphysical realism, 182
Metaphysics, 16, 44, 67, 68, 86, 95,
    126, 130, 188, 190
  of presence, 75, 188, 190
Metropolis (Fritz Lang), 59
Mimetic curse, 10, 214, 217, 227
Mitcham, Carl, 83
Monads, 69, 183, 190–1, 194
Money, 8, 148–9, 151–7
  as a general equivalent, 153–4
  in ancient Greek culture, 148–9
  and language, 151–7
  main functions of usage, 155–6
  as a technological vector of
    arithmetic, 152
  see also Language, Symbolic Forms,
    Technology, Tools
Montrose, Louis, 200
Morphology of feelings, 191
Mumford, Lewis, 93
Musil, Robert, 210
Myth, 25, 30, 32–4, 59, 106, 119, 139,
    140, 141, 143, 163, 171, 172,
    174–6, 187, 188, 191, 219, 227,
    233, 238
  fusion of myth and technology, 176
  manipulation by technology, 59,
    119, 163, 171, 175–6
  mythical consciousness, 25, 30,
    32–4, 106, 172, 187, 191
  mythological thinking, 106, 139,
    140, 141, 143, 219, 227
  technique of myth, 59
  see also Expression, Presence

Myth of the State, The (Ernst Cassirer), 5, 6, 59, 106, 113, 118, 133, 174–6, 233, 238, 254

'Nachgeborenen, An die' (Bertolt Brecht), 161, 178
National Socialism (Nazism), 1, 59, 93, 98, 107, 129, 163, 167, 175
Natorp, Paul, 114, 115
Nature, 2, 8, 10, 17, 19, 21, 23, 25–30, 32, 34–5, 37, 38, 39, 41, 42–5, 47–9
    discovery of nature contained in all technological activity, 29
    discovery of nature as disclosure, 30
    as an independent and characteristic structure, 29
    technological activity and the acknowledging of nature as independent being, 29
    theoretical knowledge of nature, 42–45
    see also Knowledge, Reality
Neher, Allister, 218
Neo-Kantianism, 7, 66, 78, 92–3, 99, 114–16, 244
Nietzsche, Friedrich, 7, 117, 127, 242

Olschki, Leonardo, 43
'On Basis Phenomena' (Ernst Cassirer), 68, 72, 74, 78, 79
Oral tradition, 148, 152, 168
Order of Discourse, The (Michel Foucault), 181
Organ-projection, 8, 10, 37–9, 147–9, 150–1, 214, 220, 227
    see also Kapp, Ernst
Organic efficacy, 37, 73
Organic existence, 38, 39, 103
Oz, Amos, 199, 200, 201, 208–9

Palahniuk, Chuck, 210
Paleo-anthropology, 8, 144
Panofsky, Erwin, 54
Paradise, 38, 82, 216, 247
    of pure organic existence, 38, 82, 247
Pech Merle cave, 219, 222–6, 227, 228

Perceptive activity, 2, 29–34, 73, 87–8, 150, 172–3, 182, 184, 185, 187, 189, 191–194
Performativity, 4, 5, 166
Phenomenology, 69, 82, 92, 95, 99, 113, 114, 116, 131, 173
Philosophical anthropology, 10, 234, 235, 240, 242, 243–4, 249–50, 251, 254
Philosophie der Technik (Friedrich Dessauer), 119
Philosophie der Technik (Ernst Kapp), 37, 119, 220
Philosophie der Technik (Eberhard Zschimmer), 119
Philosophie des Unbewußten (Eduard von Hartmann), 38
Philosophy of culture, 6, 16, 20, 113, 120, 242, 243, 250
Philosophy of mind, 7
Philosophy of science, 16, 58, 88, 92
Philosophy of Symbolic Forms, The (Ernst Cassirer), 65, 92, 99, 115, 127, 139, 141, 143, 151, 152, 172, 174, 186, 187, 203, 219
Philosophy of technology, 3, 10, 18, 19, 21, 45, 82, 83, 92, 107, 118, 119, 120, 130, 169–70, 214, 218, 228, 250
    Platonism in, 19
    see also Being of technology, Essence of technology, 'Question Concerning Technology, The', Technology
Physicalism, 57, 66, 84
Physiognomy, 72, 164, 188, 245
Plato, 19, 45–6, 48, 54, 74, 99
    the art of the craftsman, 19
    demiurge, 19
    Platonism in the philosophy of technology, 19
    see also Philosophy of technology
Poelzig, Hans, 56
Poetics (poiesis), 9, 43, 75, 78–80, 124, 200, 201, 205, 206, 207, 211
    as comprising the factual, 80
    general cultural poetics, 9, 200, 201, 205, 206, 207, 211
    poetic infinity, 78–80

Poetics (poiesis) – *continued*
  the politics of, 205, 207
  specific literary poetic, 9, 206, 207, 211
  see also Creation, Formation, Practice (Praxis), Theory (theoria), Work
Polarity, 17, 252
  genuine polarity, 17
  see also Auseinandersetzung
Postphenomenology, 82
  see also Ihde, Don
Poststructuralism, 1, 2, 7, 162, 165, 188
Practice (praxis), 4, 9, 16, 43, 28, 74, 75, 78, 88, 130, 164–5, 171, 178, 203–4, 212, 220–1, 228
  collective practice, 171
  critical practice, 164–5, 178
  literary practice, 9, 212
  science as practice, 88
  versus theoria and/or techne and/or poiesis, 4, 16, 43, 75, 78, 88
Pragmatism, 4, 16, 69, 77, 78, 88, 100, 103, 170, 203, 206, 207, 210, 212
Presence, 9, 28, 75, 181–94
  Cassirer's concept of, 190
  metaphysics of presence, 75
  presentative phenomena in language, 188
  as prior to symbolic integration, 191
  Production of Presence (Hans Ulrich Gumbrecht), 188
  rehabilitation of concept, 189
  unity of 'representation' and 'presence', 186
  see also Expression, Gumbrecht, Metaphysics
Primatology, 8, 144
Problem of Knowledge, The (Ernst Cassirer), 184
Production of Presence (Hans Ulrich Gumbrecht), 188

'Question Concerning Technology, The' (Martin Heidegger), 8, 65, 114, 118–19, 125, 129, 130
  see also Being of technology, Concept of being, Essence of technology, Philosophy of technology

Question of being, 18
  versus question of validity and question of meaning, 18
Question of meaning, 18, 20
  versus question of being and question of validity, 18
  versus question of value, 20
Question of validity, 17–18
  critical consciousness and justification, 17
  philosophical self-reflection, 17
  purpose and legitimacy of technology, 18
  versus question of being and question of meaning, 18
  quid juris, 17
  see also Self-reflection, Technology

Radio, 8, 55–9, 161–78
  genre of radio play, 161, 170
  as part of the National Socialistic propaganda apparatus, 163
  social effects of radio mediation, 161
  see also Mediation, Technology, Tool, Voice
Rancière, Jacques, 165, 176, 205
Rathenau, Walther, 48, 102, 104–5, 106, 251
Realism, 66, 75, 85–9, 182, 202, 209, 215, 217
  anti-realism, 66
  metaphysical objectivism, 66
  metaphysical realism, 182
  transformational realism, 66, 85–9
Reality, 2, 9–10, 16, 23, 24, 26, 30, 34, 43, 44, 46, 65, 68, 70–1, 77, 80–1, 86, 88, 101, 116, 129, 172–3, 182–5, 187, 191, 199–213, 214, 216, 218, 226, 227, 242, 245, 247, 251, 253
  correspondence with reality, 43, 65
  cultural images of, 9, 199–213
  formability of reality, 30, 201
  growth of reality, 44
  making of reality, 23, 71
  separation of 'I' from 'reality', 26
  see also Knowledge, Nature
Recki, Birgit, 170
Reichenbach, Hans, 56–7, 169

Representation, 2, 9, 23, 32, 85–6, 97,
100, 126, 131, 177, 181–94, 199,
201–2, 204, 207, 208
internal representations, 85
primal function of, 186
symbols not representations, 2
unity of 'presence' and
'representation', 186
see also Difference, Formation,
Literature, Mediation, Work
Representation theory, 2, 182, 202
see also Copy theory
Rhythm, 35, 39–40, 46, 83, 103, 123,
192
see also Bücher, Karl
Rickert, Heinrich, 99, 113, 114, 115
Rohbeck, Johannes, 3, 5
Rousseau, Jean-Jacques, 20–1, 47, 100,
147, 207, 247
Ruin, Hans, 7, 113
Rushdie, Salman, 201, 208, 210, 212
Ruttmann, Walter, 59

Satanic Verses, The (Salman Rushdie),
201
Sautuola, Sanz de, 215
Scepticism, 75, 98, 162–6, 169, 171, 178
Schefer, Jean Louis, 216
Schemata, 76, 123, 203–4
Schematism, 116, 117, 203
Schiller, Johann, Christoph Friedrich
von, 36, 79, 102, 117, 125, 166
art as creator of humanity, 36
ludic drive (*Spieltrieb*), 36, 102
material drive, 36
natural drive, 36
play, 36
Schmitt, Carl, 93
Science criticism, 81–3
Science and technology studies, 7, 82
Scheler, Max, 92, 233–4, 237
Schlenstedt, Dieter, 189
Schopenhauer, Arthur, 38
Seel, Martin, 193
Sein und Zeit (Martin Heidegger), 101,
114, 115, 127
Self-knowledge, 8, 34, 37–8, 73, 122,
149–51, 222, 234–5, 237, 242,
243, 246–7, 249

crisis in self-knowledge, 234, 242,
243
related to organ-projection, 37–8,
149–51, 151
technical activity as anticipation of
self-knowledge, 149–51
see also Knowledge, Self-reflection
Self-reflection, 17, 68, 70, 167
see also Question of validity, Self-
knowledge
Semiosis, 140
Shimony, Abner, 86–7
Sign theory, 2, 3, 24, 140–3, 146, 148,
153–4, 155, 172, 177, 186–9, 190
Simmel, Georg, 41, 102, 103, 104,
122, 169
tragedy of modern culture, 41, 103,
104, 122, 169
Skidelsky, Edward, 105
Sleepwalkers, The (Hermann Broch),
209
Socialism, 55, 119
Sohn-Rethel, Alfred, 151
Speech-act theory, 189
Spengler, Oswald, 93, 119–20, 176
Spinoza, Baruch, 21, 244
Stjernfelt, Frederik, 7, 92
Substance and Function (Ernst Cassirer),
67, 75, 100, 184
Symbol theory, 2, 186–9, 190
Symbolic drift, 143, 145, 152
Symbolic forms, 2–3, 5, 6, 10, 17, 66,
75, 98–102, 139–57, 156, 157,
162, 172, 173, 176–7, 182–3, 189,
190–1, 202–6, 207, 214, 218–19,
220–2, 226–7, 243–9, 251–4
gradual differentiation of, 156, 173,
191
historicity of, 177, 189, 203
as organs, 10, 222
symbolic forms and cultural
poetics, 202–6
symbolic forms as instituting
particular image worlds, 75
technology cannot be placed next
to other areas, 17
technology as symbolic form, 6,
98–102, 139–57
as tools, 3, 66

Symbolic forms – *continued*
  see also Art, Creation, Formation,
  Knowledge, Language, Mediation,
  Poetics (poiesis), Symbolic
  function
Symbolic function, 2, 67, 71, 186,
  228
  see also Symbolic forms
Symbolic pregnance, 182, 186, 192
Symposium (Plato), 45

Tale of Love and Darkness, A (Amos Oz),
  199
Technical activity, 8, 43, 54, 122,
  139–57, 222
  animal technical activity, 8, 144–6
  conscious and unconscious, 149–50
  human technical activity, 8, 144–6
  neutralization of the expressive
    dimension, 141
  relationship to language and
    mythological thought, 140–3
  and self-knowledge, 149–50, 222
  as a symbolic form, 139–57
Technological creation, 18, 34, 43, 46,
  49, 151
Technological efficacy, 18, 20, 28,
  36–8, 70, 221, 246
  condition of possibility of, 18
  as contrasted to technological
    objects, 20
Technological formation, 18, 70, 162
  condition of possibility of, 18
Technological objects, 20, 221
  see also Formation, Work
Technological mediation, 7, 82, 104
Technological thinking, 39
  see also Thinking
Technological will, 25, 42
  versus magical desire, 25
Technological work, 18, 35, 39, 40,
  43–7, 103
Technologization, 94, 96–7, 99, 102,
  105–6, 114, 118
Technology
  achievement of, 34
  as applied natural science, 19
  modern technology, 5–6, 41, 126,
    134, 162, 166, 169

nano-bio-info-cogno (NBIC)
  technologies, 10
NBIC convergence, 10
and the possible, 80–1
as primary medium, 3
purpose and legitimacy of
  technology, 18
relationship to language, 142, 166,
  167, 221
as tool of the mind, 3, 39, 89
true definition of technology is
  genetic, 100
of writing and reading, 152
see also Being of technology, Essence
  of technology, Knowledge,
  Mediation, Philosophy
  of technology, 'Question
  Concerning Technology, The',
  Question of validity, Symbolic
  forms, Technological efficacy,
  Technological formation, Tools
Testart, Alain, 153
Theory (Theoria), 3, 4, 16, 23, 43,
  74–5, 88, 165, 170, 173, 176, 177,
  186–7
  instrumental and theoretical
    meaning of logos, 23
  performative aspect of, 4, 170,
    176
  versus practice (praxis) and/or
    techne and/or poiesis, 4, 16, 43,
    74–5, 88
  theoretical work performed by
    instruments, 3
  see also Aristotle, Formation,
    Instruments, Knowledge, Logos,
    Practice (praxis), Tools
Theory of representation, 181
  see also Copy theory,
    Representation theory
Thinking, 7, 10, 18, 24, 27, 30, 31, 35,
  42, 49, 65–89, 130, 132, 190, 200,
  204, 227, 245, 250, 253
  anticipatory thinking, 31, 35, 80
  causal thinking, 10, 27, 250
  and doing united, 16, 24, 42
  line of sight, 31
  magical thinking, 253
  materialistic ways of thinking, 18

mediacy that makes thinking possible, 7, 30
mythological thinking, 227, 245
practical thinking, 49
reflexive thinking, 132
speaking and thinking, 24
technics of thinking, 65–89
technology and fore-seeing and fore-thought, 31
see also Formation, Logos, Mediation, Technological thinking, Theory (Theoria), Work
Third Reich, 5, 163, 175, 176
propaganda machinery of, 175
Time, 31, 45, 80, 97, 242
technology and fore-seeing and fore-thought, 31, 45, 80
technology and future-directedness, 45, 242
see also Technology, Thinking
Tool-making, 80, 142–7
mythical aspects of, 155
relationship to language, 142–7
tool-making animal, 24, 66, 80
Tools, 19, 22–4, 32, 38, 39
axes, 32
chisels, 38
drills, 38
hammers, 32, 33
instrumental and theoretical meaning of logos, 23
looms, 19
sewing machines, 39
steel mills, 39
tongs, 38
tools of the mind, 23
Tomasello, Michael, 146
'Totale Mobilmachung, Die' (Ernst Jünger), 93
Transcendental idealism, 121, 127, 184
Transcendental phenomenology, 69–70
Transformation, 3, 18, 25, 33, 39, 66, 74, 119, 121, 168, 174–5, 178, 209, 218, 221, 240
creation and transformation, 121, 209
formation and qualitative transformation, 18, 218

metamorphosis, 18
mind as transformation of life, 74, 84
transformational realism, 66, 85–9
see also Formation, Mediation, Work
Transposition, 9, 10, 208–13
versus creation and exposition, 208–13
Tygstrup, Frederik, 9, 199

'Über den Schmerz' (Ernst Jünger), 93
Uexküll, Jakob von, 72, 243, 244
Umwelt, 72
'Unity of Science, The' (Ernst Cassirer), 57
Utne, Lise, 213

Value theory, 16
Vampiric forces, 35, 73, 122, 235
Verfremdungs-Effekt, 164, 165, 177
see also Alienation effects, Bertolt Brecht, Distancing strategies
Vitalism, 92, 99, 102
Voice, 8–9, 56, 93, 161–78, 208
expressivity of, 164, 170, 171
listening reception of, 161–78
in literature, 208

Was ist Radio? (Hans Reichenbach), 169
Weber, Max, 209
Weimar Republic, 1, 9, 107, 161, 164, 167, 168, 170, 177
new mass media, 161, 167
radio discourse of, 164, 168
Weiss, Dennis M., 10, 233
Welsch, Wolfgang, 163–4, 175
Windelband, Wilhelm, 115
Work, 3, 4, 18, 19, 22, 33–4, 35, 39–40, 43–8, 49, 54, 68–9, 73, 74, 75, 78–80, 84–5, 98, 100, 103–5, 140, 142, 163, 173, 205, 208–9, 218, 224, 228, 244, 248
manual work, 54
mechanized work, 48, 105
the region of the effected and created, 18
settling-of-action-into-work, 78
solidarity of work, 49
technological works, 18
work as basis phenomenon, 74

Work – *continued*
work and poetic infinity, 78–80
works of art (artwork), 46, 80, 104,
125, 163–4, 172, 190–1, 205,
208–9, 221
work-aspect compared with I-aspect
and action-aspect, 68–9, 78,
84–5

see also Auseinandersetzung,
Basis phenomena, Difference,
Formation, Mediation
Worker: Dominion and Gestalt, The
(Ernst Jünger), 7, 92–108, 120

Žižek, Slavoj, 205
Zschimmer, Eberhard, 119